Mathematics Workbook

AGE 9–11

Mental Arithmetic

David E Hanson

GALORE PARK
AN HACHETTE UK COMPANY

About the author

David Hanson has over 40 years' experience of teaching and has been leader of the Independent Schools Examinations Board (ISEB) 11+ Maths setting team, a member of the ISEB 13+ Maths setting team and a member of the ISEB Editorial Endorsement Committee.

Every effort has been made to trace all copyright holders, but if any have been inadvertently overlooked the publishers will be pleased to make the necessary arrangements at the first opportunity.

Although every effort has been made to ensure that website addresses are correct at time of going to press, Galore Park cannot be held responsible for the content of any website mentioned in this book. It is sometimes possible to find a relocated web page by typing in the address of the home page for a website in the URL window of your browser.

Hachette UK's policy is to use papers that are natural, renewable and recyclable products and made from wood grown in well-managed forests and other controlled sources. The logging and manufacturing processes are expected to conform to the environmental regulations of the country of origin.

Orders: Teachers please contact Bookpoint Ltd, 130 Park Drive, Milton Park, Abingdon, Oxon OX14 4SE. Telephone: (44) 01235 400555. Email primary@bookpoint.co.uk. Lines are open from 9 a.m. to 5 p.m., Monday to Saturday, with a 24-hour message answering service

Parents, Tutors please call: 020 3122 6405 (Monday to Friday, 9:30 a.m. to 4.30 p.m.). Email: parentenquiries@galorepark.co.uk

Visit our website at www.galorepark.co.uk for details of other revision guides for Common Entrance, examination papers and Galore Park publications.

ISBN: 978 1 471829 62 8

Text copyright © David Hanson 2014
The right of David Hanson to be identified as the author of this Work has been asserted by him in accordance with sections 77 and 78 of the Copyright, Designs and Patents Act 1988.

First published by Galore Park Publishing Ltd
An Hachette UK company
Carmelite House
50 Victoria Embankment
London EC4Y 0DZ

www.galorepark.co.uk

Impression number   10 9 8
2022

Typeset in India
Printed in Great Britain by Ashford Colour Press Ltd
Illustrations by Aptara, Inc.

A catalogue record for this title is available from the British Library.

Contents and test results

	Page	Completed	Marks	%	Time taken
Test 1	7				
Test 2	8				
Test 3	9				
Test 4	10				
Test 5	11				
Test 6	12				
Test 7	13				
Test 8	14				
Test 9	15				
Test 10	16				
Test 11	17				
Test 12	18				
Test 13	19				
Test 14	20				
Test 15	21				
Test 16	22				
Test 17	23				
Test 18	24				
Test 19	25				
Test 20	26				
Test 21	27				
Test 22	28				
Test 23	29				
Test 24	30				
Test 25	31				
Test 26	32				
Test 27	33				
Test 28	34				
Test 29	35				
Test 30	36				
Test 31	37				
Test 32	38				
Test 33	39				
Test 34	40				

	Page	Completed	Marks	%	Time taken
Test 35	41				
Test 36	42				
Test 37	43				
Test 38	44				
Test 39	45				
Test 40	46				
Test 41	47				
Test 42	48				
Test 43	49				
Test 44	50				
Test 45	51				
Test 46	52				
Test 47	53				
Test 48	54				
Test 49	55				
Test 50	56				

Answers to all the questions in this book can be found in the pull-out section in the middle.

Introduction

- **Arithmetic**: the branch of mathematics dealing with the properties and manipulation of numbers.
- **Mental arithmetic**: arithmetical calculations performed in the mind, without writing figures or using a calculator.

This book consists of 50 tests. Each test:

- is printed in two columns on one page
- consists of 20 questions
- has a total of 20 marks.

The tests:

- cover a full range of arithmetic skills
- follow the same general pattern
- feature a gradual increase in difficulty
- are designed to:
  - provide practice in recalling number facts and procedures
  - facilitate the identification of weak areas
  - encourage working quickly and accurately
  - facilitate the monitoring of progress
  - build pupil confidence.

Odd-numbered tests start with 10 straightforward questions which act as reminders of 10 basic strategies. At first there are hints about possible ways forward for each question but these become shorter and less descriptive as the book progresses.

The tests can be used in two main ways:

- Complete the test, as quickly as possible, recording the time taken.
- Do as much as possible in a fixed time.

All the questions:

- are to be tackled entirely in the head, without
  - doing any working
  - any measuring instruments
  - a calculator.
- require a single response only
- are answered on the line to the right or underneath of the questions; units are given where appropriate.

All questions require a degree of mental activity and there are no questions such as 'name this shape'.

Answers to the questions can be pulled out of the middle of the book.

Where appropriate, answers involving fractions should be given in their simplest form.

Test 1

For all of the questions in this test, do the calculation entirely in your head with no written 'working' and just write down the answer.

In questions 1 to 10 you are reminded of 10 useful strategies which may help you in later questions.

1 3 × 29 _____

   ✳ 3 × 30 and then subtract 3

2 £12.40 ÷ 4 £ _____

   ✳ Divide by 2 and then by 2 again

3 48 × 5 _____

   ✳ Multiply by 10 then divide by 2

4 65 – 49 _____

   ✳ Subtract 50 then add 1

5 204 ÷ 6 _____

   ✳ Divide by 2, then by 3

6 0.7 × 6 _____

   ✳ You know that 7 × 6 = 42

7 $\frac{3}{5}$ of 30 _____

   ✳ Find $\frac{1}{5}$ then multiply by 3

8 17 + 28 + 33 _____

   ✳ 17 + 33 = 50 first

9 3.9 × 2 _____

   ✳ The result must be about 8

10 23 × 8 _____

   ✳ The units digit of the result is 4

11 What temperature is 7 degrees lower than 4 °C?

   _____°C

12 Round 5.47 to 1 decimal place. _____

13 Write $\frac{6}{8}$ in its simplest form. _____

14 What fraction of the strip of squares is red?

   _____

15 Given that 47 × 34 = 1598, what is 47 × 33?

   _____

16 What is the remainder when 100 is divided by 7?

   _____

17 By how much is 4 × 13 greater than 3 × 14?

   _____

18 Joanna buys 2 bottles of juice priced at 95p each and a packet of crisps priced at 89p.

   How much change will she receive from a £5 note?

   £ _____

19 Sean is thinking of a number and gives the following clues:

   "My number is:
   • less than 30
   • a multiple of 7
   • 1 more than a prime number."

   What number is Sean thinking of?

   _____

20 On this tower of bricks, the number on each brick is the sum of the numbers on the two bricks supporting it.

   What number is on the top brick? _____

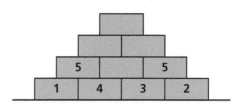

Test 2

For all of the questions in this test, do the calculation entirely in your head with no written 'working' and just write down the answer.

1 Round 4745 to the nearest 100 _____

2 What is 20% of 40 kg? _____ kg

3 George and Mildred share 12 sweets in the ratio 1:2

 How many sweets does George get? _____

4 Angus has the five number cards below.

 What is the smallest 5-digit even number he can make by placing all five cards side by side? _____

5 When Fay thought of a number, multiplied it by 4 and then added 5, the result was 17

 What number did Fay think of? _____

6 If $a = 5$, $b = 4$ and $c = 3$, what is the value of $ab + c$? _____

7 What is the next number in this sequence?

 1, 3, 6, 10, 15, _____

8 Tim has a bag of sweets. He gives 4 to Miranda and 5 each to Sam and Mary. Tim tips the rest of the sweets onto the table and counts 11

 How many sweets were in the bag at the start? _____

9 Calculate the perimeter of the rectangle.

 _____ cm

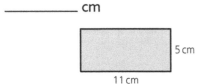

10 What is the reading on the scale? _____

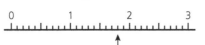

11 Two angles of a triangle are 60° and 110°.

 What size is the third angle? _____°

12 What is the volume of a cuboid measuring 5 cm by 4 cm by 3 cm?

 _____ cm³

13 A disco started at 18:45 and lasted 2 hours and 30 minutes.

 At what time did it end? _____ : _____

Questions 14 to 16 refer to this list of marks one pupil gained in 5 tests.

10 6 9 5 10

14 What is the mode? _____

15 What is the median? _____

16 What is the mean? _____

17 The Venn diagram shows some information about children in Year 6

 How many boys play the piano? _____

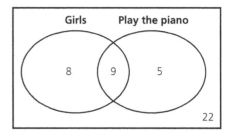

18 Which number between 65 and 70 is divisible exactly by 6?

 _____

19 The product of two integers (whole numbers) both less than 20 is 66

 What is the sum of the numbers? _____

20 Jennie has 20 buttons. 8 are black and half of the others are red. The rest are white.

 How many buttons are white? _____

Test 3

For all of the questions in this test, do the calculation entirely in your head with no written 'working' and just write down the answer.

In questions 1 to 10 you are reminded of 10 useful strategies which may help you in later questions.

1 4 × 19 _____

 ✷ *4 × 20 and then subtract 4*

2 £8.40 × 4 £ _____

 ✷ *Double and then double again*

3 130 ÷ 5 _____

 ✷ *Divide by 10 then multiply by 2*

4 73 + 38 _____

 ✷ *Add 40 then subtract 2*

5 336 ÷ 12 _____

 ✷ *Divide by 3, then by 4*

6 0.8 × 5 _____

 ✷ *You know that 8 × 5 = 40*

7 $\frac{3}{4}$ of 48 _____

 ✷ *Find $\frac{1}{4}$ then multiply by 3*

8 25 + 17 + 55 _____

 ✷ *25 + 55 = 80 first*

9 5.1 × 8 _____

 ✷ *The result must be about 40*

10 17 × 6 _____

 ✷ *The units digit of the result is 2*

11 What temperature is 4 degrees higher than ⁻2°C?

_____ °C

12 Round 35.8 to 1 significant figure. _____

13 Write $\frac{5}{4}$ as a mixed number. _____

14 What proportion of the strip of squares is shaded?

_____

15 Given that 7134 ÷ 58 = 123, what is 7134 ÷ 29?

_____

16 What is the remainder when 60 is divided by 8?

_____

17 By how much is 6 × 15 greater than 5 × 16?

_____

18 Gary buys 3 cans of cola priced at 55p each and a chocolate bar priced at 78p.

How much change will he receive from two £2 coins?

£ _____

19 Melissa is thinking of a number and gives the following clues:

"My number is:

• between 20 and 40
• a multiple of 3
• 1 less than a square number."

What number is Melissa thinking of?

_____

20 On this tower of bricks, the number on each brick is the sum of the numbers on the two bricks supporting it.

What number is on the top brick? _____

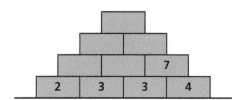

Test 4

For all of the questions in this test, do the calculation entirely in your head with no written 'working' and just write down the answer.

1 Round 40 800 to the nearest 1000

_____

2 What is 60% of 40 ml? _____ ml

3 Philip and Mary share £2 in the ratio 3 : 2

How much does Mary get? _____

4 Harry has the six number cards below.

What is the number nearest to 1000 he can make by placing some of the cards side by side? _____

5 When Kirsten thought of a number, added 1 and then multiplied by 3, the result was 9

What number did Kirsten think of? _____

6 If $a = 3$, $b = 2$ and $c = 8$, what is the value of $a - b + c$?

_____

7 What is the next number in this sequence?

1, 2, 4, 7, 11, _____

8 Isobel has a packet of sweets. She shares them with her sister. When they have each eaten 7 sweets there is 1 sweet left.

How many sweets were in the packet at the start?

_____

9 Calculate the area of the rectangle.

_____cm²

[rectangle labelled 2 cm on the right side and 12 cm below]

10 What is the reading on the scale? _____

[number line from 0 to 5 with an arrow pointing just below 2]

11 Two angles of a triangle are 40° and 70°.

What size is the third angle? _____°

12 What is the volume of a cuboid measuring 8 cm by 5 cm by 4 cm?

_____cm³

13 A film started at 7.50 p.m. and ended at 10.05 p.m. How long did the film last?

_____ hours _____ minutes

Questions 14 to 16 refer to this list of the marks one pupil gained in 8 tests.

8 6 9 7 6 8 7 6

14 What is the mode? _____

15 What is the median? _____

16 What is the range? _____

17 The Carroll diagram shows some information about children in a class.

What percentage of the children in the class wear glasses? _____%

	Boys	Girls
Do not wear glasses	7	6
Wear glasses	4	3

18 What is the smallest number greater than 100 that is divisible exactly by 3?

_____

19 The sum of two integers (whole numbers) is 12 and the difference between the integers is 4

What is the product of the numbers?

_____

20 Paige has 24 sweets. A quarter are mints and a third of the others are toffees. The rest are chocolates.

How many chocolates are there? _____

Test 5

For all of the questions in this test, do the calculation entirely in your head with no written 'working' and just write down the answer.

In questions 1 to 10 you are reminded of 10 useful strategies which may help you in later questions.

1   11 × 19 _____

 ✳ 10 × 19 and then add 19

2   128 ÷ 8 _____

 ✳ Divide by 2, by 2 and then by 2 again

3   123 × 5 _____

 ✳ Multiply by 10 then divide by 2

4   85 – 38 _____

 ✳ Subtract 40 then add 2

5   330 ÷ 15 _____

 ✳ Divide by 3, then by 5

6   1.3 × 4 _____

 ✳ You know that 13 × 4 = 52

7   $\frac{2}{3}$ of 72 _____

 ✳ Find $\frac{1}{3}$ then multiply by 2

8   4.8 + 3.5 + 1.2 _____

 ✳ 4.8 + 1.2 = 6.0 first

9   5 × 19 _____

 ✳ The result must be about 100

10 13 × 9 _____

 ✳ The units digit of the result is 7

11 What temperature is 11 degrees lower than 7 °C?

 _____ °C

12 Round 3.494 to 2 decimal places. _____

13 Write $\frac{8}{12}$ in its simplest form. _____

14 What fraction (in its simplest form) of the shape is red?

 _____

15 Given that 35 × 62 = 2170, what is 34 × 62?

 _____

16 What is the remainder when 80 is divided by 6?

 _____

17 By how much is 4^2 greater than 3^2?

 _____

18 Lacey buys 2 bottles of lemonade priced at £1.05 each and 2 packets of crisps priced at 75p.

 How much change will she receive from a £5 note? £ _____

19 Joe is thinking of a number and gives the following clues:

 "My number is:
 • between 10 and 30
 • 1 less than a square number
 • 1 more than a multiple of 7"

 What number is Joe thinking of? _____

20 On this tower of bricks, the number on each brick is the sum of the numbers on the two bricks supporting it.

 What number is on the brick marked with a star (*)? _____

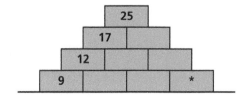

Test 6

For all of the questions in this test, do the calculation entirely in your head with no written 'working' and just write down the answer.

1 Round 405 600 to the nearest 10 000

_____

2 What is 15% of £10.00?   £ _____

3 Sienna and Corey share 40 conkers in the ratio 1:3

How many conkers does Corey get?

_____

4 Libby has the four number cards below.

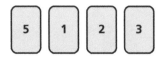

What is the smallest 4-digit even number she can make by placing all four cards side by side? _____

5 When Joel thought of a number, multiplied it by 3 and then subtracted 4, the result was 17

What number did Joel think of? _____

6 If $a = 3$, $b = 6$ and $c = ^-3$, what is the value of $ab + c$? _____

7 What is the next number in this sequence?

1, 3, 7, 15, 31, _____

8 Reuben has a collection of stamps. He gives a third of the stamps to his little brother Archie. Archie received 44 stamps.

How many stamps does Reuben have now?

_____

9 Calculate the perimeter of the rectangle. _____ cm

8.5 cm
2.5 cm

10 What is the reading on the scale? _____

11 Two angles of a triangle are 55° and 105°.

What size is the third angle? _____ °

12 What is the volume of a cuboid measuring 2 cm by 8 cm by 2 cm?

_____ cm³

13 A $2\frac{1}{2}$ hour film ended at 21:10

At what time did it start? _____:_____

Questions 14 to 16 refer to this list of the marks one pupil gained in 10 tests.

8 7 6 8 7 8 5 9 8 9

14 What is the mode? _____

15 What is the median? _____

16 What is the mean? _____

17 The Venn diagram shows some information about people in a sports club.

How many people are in the club?

_____

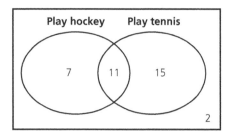

18 Which number between 80 and 90 is divisible exactly by 14?

_____

19 The product of two integers (whole numbers) both less than 10 is 35

What is the sum of the numbers? _____

20 A farmer has 120 animals. 40 are cows and $\frac{3}{4}$ of the others are sheep. The rest are pigs.

How many pigs are there? _____

Test 7

For all of the questions in this test, do the calculation entirely in your head with no written 'working' and just write down the answer.

In questions 1 to 10 you are reminded of 10 useful strategies which may help you in later questions.

1 29 × 11 _____

   ✳ 30 × 11 and then subtract 11

2 £2.20 × 8 £ _____

   ✳ Double, double again and then double again

3 220 ÷ 5 _____

   ✳ Divide by 10 then multiply by 2

4 145 + 59 _____

   ✳ Add 60 then subtract 1

5 234 ÷ 18 _____

   ✳ Divide by 3, then by 6

6 0.7 × 8 _____

   ✳ You know that 7 × 8 = 56

7 $\frac{2}{5}$ of 70 _____

   ✳ Find $\frac{1}{5}$ then multiply by 2

8 23 + 19 + 57 + 21 _____

   ✳ 23 + 57 and 19 + 21 first

9 4.9 × 4 _____

   ✳ The result must be about 20

10 14 × 5 _____

   ✳ The units digit of the result is 0

11 What temperature is 6 degrees higher than ⁻3 °C?

   _____°C

12 Round 44.9 to 1 significant figure. _____

13 Write $3\frac{1}{3}$ as an improper fraction. _____

14 What proportion of the strip of squares is red?

   _____

   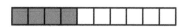

15 Given that 1176 ÷ 28 = 42, what is 1176 ÷ 56?

   _____

16 What is the remainder when 50 is divided by 9?

   _____

17 By how much is 7 × 12 greater than 2 × 17?

   _____

18 Ellis buys 2 cans of cola priced at 59p each and 2 muffins priced at £1.11 each.

   How much change will he receive from a £10 note?

   £ _____

19 Maisie is thinking of a number and gives the following clues:

   "My number is:
   • between 30 and 60
   • a multiple of 7
   • 1 less than a prime number."

   What number is Maisie thinking of?

   _____

20 On this tower of bricks, the number on a brick is the **product** of the numbers on the two bricks supporting it.

   What number is on the top brick? _____

   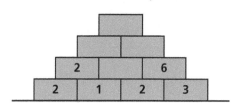

Test 8

For all of the questions in this test, do the calculation entirely in your head with no written 'working' and just write down the answer.

1 Round 49 819 to the nearest 1000

_____

2 What is 25% of 240 kg? _____ kg

3 Anna and Tia share £4.50 in the ratio 2 : 1

How much does Tia get? £ _____

4 Blake has the five number cards below.

What is the number nearest to 500 he can make by placing some of the cards side by side? _____

5 When Rosie thought of a number, added 3 and then multiplied by 4, the result was 24

What number did Rosie think of? _____

6 If $a = 1$, $b = 3$ and $c = 4$, what is the value of $b - a + c$? _____

7 What is the next number in this sequence?

1, 2, 5, 14, 41, _____

8 Layla has a bag of chews. She shares them with her friends, Maya and Niamh. When they have each eaten 4 chews there are 2 chews left.

How many chews were in the bag at the start? _____

9 Calculate the area of the rectangle.
_____ cm²

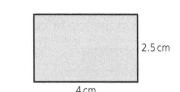

2.5 cm

4 cm

10 What is the reading on the scale? _____

```
1        2        3        4
|⌊ꞁꞁꞁꞁꞁ|ꞁꞁꞁꞁꞁ|ꞁꞁꞁꞁꞁ|ꞁꞁꞁꞁꞁ|
         ↑
```

11 Two angles of a triangle are 63° and 89°.

What size is the third angle? _____ °

12 What is the volume of a cuboid measuring 7 cm by 2 cm by 5 cm?

_____ cm³

13 A concert started at 6.45 p.m. and ended at 9.25 p.m. How long did the concert last?

_____ hours _____ minutes

Questions 14 to 16 refer to this list of the marks one pupil gained in 5 tests.

9 7 9 5 10

14 What is the range? _____

15 What is the median? _____

16 What is the mean? _____

17 The Carroll diagram shows some information about children in a class.

What percentage of the children in the class plays chess? _____ %

	Boys	Girls
Do not play chess	3	2
Play chess	7	8

18 What is the smallest number greater than 90 that is divisible exactly by 4?

_____

19 The sum of two integers (whole numbers) is 17 and the difference between the integers is 7

What is the product of the numbers?

_____

20 Zara has 20 British coins. A quarter are 20p coins and the rest are 50p coins.

What is the total value of Zara's coins?

£ _____

Test 9

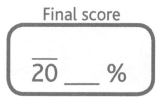
For all of the questions in this test, do the calculation entirely in your head with no written 'working' and just write down the answer.

In questions 1 to 10 you are reminded of 10 useful strategies which may help you in later questions.

1 21 × 11 _____

   ✱ *Take easier steps when you multiply*

2 224 ÷ 8 _____

   ✱ *Double or halve*

3 48 × 5 _____

   ✱ *Use your 10s and 2s*

4 93 − 29 _____

   ✱ *Take easier steps when you subtract*

5 396 ÷ 12 _____

   ✱ *Use factors to divide*

6 0.8 × 7 _____

   ✱ *Use known facts*

7 $\frac{2}{3}$ of 42 _____

   ✱ *Use a step-by-step approach*

8 8.3 + 4.9 + 1.7 _____

   ✱ *Group when you add*

9 6 × 19 _____

   ✱ *Approximate the result*

10 12 × 13 _____

   ✱ *Check using the units digit of the result*

11 What temperature is 9 degrees lower than 8 °C?

   _____ °C

12 Round 47.47 to 1 decimal place. _____

13 Write $\frac{12}{18}$ in its simplest form.

   _____

14 What percentage of the shape is red?

   _____ %

15 Given that 54 × 86 = 4644, what is 27 × 86?

   _____

16 What is the remainder when 100 is divided by 9?

   _____

17 By how much is 5^2 greater than 4^2?

   _____

18 Lexi buys 2 cans of cola priced at 59p each and 2 packets of crisps priced at 61p each and a chocolate bar costing £1.50

   How much change will she receive from a £10 note?

   £ _____

19 Jon is thinking of a 2-digit number and gives the following clues:

   "My number is:

   • less than 60
   • 1 less than a square number

   and the digits add to 12"

   What number is Jon thinking of? _____

20 On this tower of bricks, the number on each brick is the sum of the numbers on the two bricks supporting it.

   What number is on the brick marked with a star (*)?

   _____

		43	
	19		
11			
4		*	

Test 10

Final score

$\overline{20}$ ___ %

For all of the questions in this test, do the calculation entirely in your head with no written 'working' and just write down the answer.

1 Round 12 345 to the nearest 100 _____

2 What is 30% of £18.00? £ _____

3 Maddison and Jayden share 24 sweets in the ratio 1:5

How many sweets does Maddison get? _____

4 Adam has the three number cards below.

What is the difference between the largest and smallest 3-digit numbers he can make by placing all three cards side by side? _____

5 When Alisha thought of a number, multiplied it by 4 and then subtracted 2, the result was 18

What number did Alisha think of? _____

6 If $a = 4$, $b = 1$ and $c = {}^-2$, what is the value of $a - b - c$? _____

7 What is the next number in this sequence?

1, 3, 11, 43, _____

8 Ewan has a tin of foreign coins. He puts them in bags of 10 coins and there are 3 coins left over. He sells the bags of coins at £2 each and gives the money to charity. The charity receives £24

How many coins were in Ewan's tin? _____

9 Calculate the perimeter of the rectangle. _____ cm

☐ 0.8 cm
13.2 cm

10 What is the reading on the scale? _____

0   1

11 Two angles of a triangle are 83° and 23°.

What size is the third angle? _____ °

12 What is the volume of a cuboid measuring 5 cm by 4 cm by 3 cm?

_____ cm³

13 A $1\frac{3}{4}$ hour concert ended at 20:40

At what time did it start? _____:_____

Questions 14 to 16 refer to this list of the marks one pupil gained in 10 tests.

8 7 8 5 9 7 6 8 9 6

14 What is the mode? _____

15 What is the median? _____

16 What is the mean? _____

17 The Venn diagram shows some information about people on a bus.

How many females were on the bus? _____

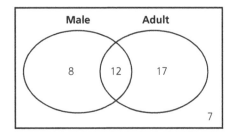

18 Which number between 100 and 105 is divisible exactly by 6?

_____

19 The sum of two integers (whole numbers) both less than 30 is 50 and the difference between them is 8

What is the larger number? _____

20 A club has 67 members. There are 34 junior members and a third of the senior members are men.

How many of the senior members are women?

_____

16

Test 11

For all of the questions in this test, do the calculation entirely in your head with no written 'working' and just write down the answer.

In questions 1 to 10 you are reminded of 10 useful strategies which may help you in later questions.

1 61×21 _____

   ✳ *60 × 21 and then add 21*

2 £3.45 × 4 £ _____

   ✳ *Double and then double again*

3 440 ÷ 5 _____

   ✳ *Divide by 10 then multiply by 2*

4 77 + 69 _____

   ✳ *Add 70 then subtract 1*

5 225 ÷ 15 _____

   ✳ *Divide by 3, then by 5*

6 0.9 × 12 _____

   ✳ *You know that 9 × 12 = 108*

7 $\frac{3}{4}$ of 76 _____

   ✳ *Find $\frac{1}{4}$ then multiply by 3*

8 17 + 44 + 23 + 5 _____

   ✳ *17 + 23 first*

9 2.05 × 4 _____

   ✳ *The result must be about 8*

10 36 × 4 _____

   ✳ *The units digit of the result is 4*

11 What temperature is 2 degrees lower than ⁻3 °C?

   _____°C

12 Round 54.8 to 2 significant figures. _____

13 Write $\frac{7}{3}$ as a mixed number. _____

14 What proportion of the rectangle is blue?

   _____

15 Given that 864 ÷ 18 = 48, what is 864 ÷ 36?

   _____

16 What is the remainder when 70 is divided by 8?

   _____

17 By how much is 8 × 12 greater than 2 × 18?

   _____

18 Sofia buys 3 cans of cola priced at 61p each and 3 cupcakes priced at 49p each.

   How much change will she receive from a £5 note?

   £ _____

19 Riley is thinking of a number and gives the following clues:

   "My number is:

   • between 3 and 23
   • 1 less than a prime number
   • 1 more than a square number."

   What number is Riley thinking of? _____

20 On this tower of bricks, the number on a brick is the **product** of the numbers on the two bricks supporting it.

   What number is on the top brick? _____

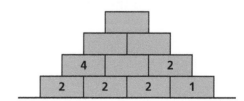

Test 12

For all of the questions in this test, do the calculation entirely in your head with no written 'working' and just write down the answer.

1 Round 101010 to the nearest 10000

_____

2 What is 40% of 5 litres? _____ litres

3 Harvey and Harrison share £20 so that Harvey gets £4 more than Harrison.

How much does Harrison get? £ _____

4 Skye has the five number cards below.

| 3 | 8 | 5 | 0 | 7 |

What is the largest multiple of 5 she can make by placing all five cards side by side?

_____

5 When Lucas thought of a number, subtracted 3 and then multiplied by 2, the result was 2

What number did Lucas think of? _____

6 If $a = 5$, $b = {}^-2$ and $c = 1$, what is the value of $a - b + c$? _____

7 What is the next number in this sequence?

1, 2, 4, 8, 16, _____

8 Henry has a tin of toffees. He shares them with his friends, Nicole and Josh. When they have each eaten 7 toffees there are 5 toffees left.

How many toffees were in the tin at the start? _____

9 Calculate the area of the rectangle.
_____ cm²

4 cm

8.5 cm

10 What is the reading on the scale? _____

0   0.5

11 Two angles of a triangle are 24° and 106°.

What size is the third angle? _____ °

12 What is the volume of a cuboid measuring 10 cm by 4 cm by 8 cm?

_____ cm³

13 Hattie left home at 11.45 a.m. and reached the bank at 1.05 p.m. How long did the journey take her?

_____ hour(s) _____ minutes

Questions 14 to 16 refer to this list of the marks one pupil gained in 5 tests.

13 17 19 11 20

14 What is the range? _____

15 What is the median? _____

16 What is the mean? _____

17 The Carroll diagram shows some information about children in a class.

What fraction of the children in the class plays football? _____

	Boys	Girls
Do not play football	5	6
Play football	7	4

18 What is the largest number less than 100 that is divisible exactly by 7?

_____

19 The sum of two integers (whole numbers) is 60 and the difference between the integers is 12

What is the larger number? _____

20 Kara has a collection of 48 stamps. 16 are American, half the others are from mainland Europe and the remainder are British.

How many British stamps does Kara have?

_____

Test 13

For all of the questions in this test, do the calculation entirely in your head with no written 'working' and just write down the answer.

In questions 1 to 10 you are reminded of 10 useful strategies which may help you in later questions.

1   19 × 13 _____

   ∗ Take easier steps when you multiply

2   224 ÷ 8 _____

   ∗ Double or halve

3   38 × 5 _____

   ∗ Use your 10s and 2s

4   81 − 47 _____

   ∗ Take easier steps when you subtract

5   378 ÷ 21 _____

   ∗ Use factors to divide

6   1.4 × 5 _____

   ∗ Use known facts

7   $\frac{3}{8}$ of 24 _____

   ∗ Use a step-by-step approach

8   4.8 + 1.7 + 1.3 + 2.2 _____

   ∗ Group when you add

9   4 × 29 _____

   ∗ Approximate the result

10 13 × 13 _____

   ∗ Check using the units digit of the result

11 What temperature is 7 degrees lower than 1°C?

   _____°C

12 Round 105.85 to 1 decimal place. _____

13 Write $\frac{20}{36}$ in its simplest form. _____

14 What fraction of the shape is blue?

   _____

15 Given that 21 × 32 = 672, what is 42 × 16?

   _____

16 What is the remainder when 200 is divided by 3?

   _____

17 By how much is 8^2 greater than 7^2?

   _____

18 Nicole buys 3 bottles of lemonade priced at 85p each and 3 sausage rolls priced at £1.15 each.

   How much change will she receive from a £10 note?

   £ _____

19 Faith is thinking of a 2-digit number and gives the following clues:

   "My number is a square number and the digits add to 10"

   What number is Faith thinking of? _____

20 On this tower of bricks, the number on each brick is the sum of the numbers on the two bricks supporting it.

   What number is on the brick marked with a star (∗)?

   _____

		56		
			30	
	11			
∗				4

Test 14

For all of the questions in this test, do the calculation entirely in your head with no written 'working' and just write down the answer.

1   Round 286 to the nearest 50 _____

2   What is 60% of 5 litres? _____ litres

3   Tarik and Ivan share 25 chocolates in the ratio 2 : 3

   How many more chocolates than Tarik does Ivan get? _____ more

4   Marcus has the four number cards below.

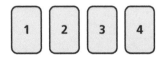

   What is the largest 2-digit multiple of 3 he can make by placing two of the cards side by side? _____

5   When Lucie thought of a number, multiplied it by 5 and then subtracted 3, the result was 32

   What number did Lucie think of? _____

6   If $a = 2$, $b = {}^-2$ and $c = {}^-3$, what is the value of $a + b - c$? _____

7   What is the next number in this sequence?

   1, 2, 5, 14, _____

8   Gabriel has a box of stamps. He puts them in bags of 50 stamps and there are 23 stamps left over. He sells the bags of stamps at £1 each and gives the money to charity. The charity receives £12

   How many stamps were in Gabriel's box?

   _____

9   Calculate the perimeter of the rectangle.
   _____ cm

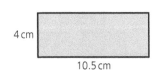

4 cm

10.5 cm

10 What is the reading on the scale? _____

0   1

11 Two angles of a triangle are 31° and 47°.

   What size is the third angle? _____°

12 What is the volume of a cuboid measuring 10 cm by 10 cm by 5 cm?

   _____ cm³

13 A $2\frac{1}{4}$ hour film ended at 22:10

   At what time did it start? _____ : _____

Questions 14 to 16 refer to this list of the marks one pupil gained in 5 tests.

18 16 19 14 18

14 What is the mode? _____

15 What is the median? _____

16 What is the mean? _____

17 The Venn diagram shows some information about people at a concert.

   How many people were at the concert?

   _____

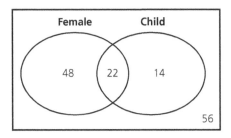

18 Which number is exactly half way between 50 and 64?

   _____

19 The sum of two integers (whole numbers) is 45 and the difference between them is 7

   What is the larger number? _____

20 A school has 148 pupils. There are 16 more boys than girls.

   How many girls are there? _____

Test 15

For all of the questions in this test, do the calculation entirely in your head with no written 'working' and just write down the answer.

In questions 1 to 10 you are reminded of 10 useful strategies which may help you in later questions.

1 41 × 22 _____

   ✳ 40 × 22 and then add 22

2 34 × 4 _____

   ✳ Double and then double again

3 240 ÷ 5 _____

   ✳ Divide by 10 then multiply by 2

4 49 + 72 _____

   ✳ Add 50 to 72 then subtract 1

5 156 ÷ 12 _____

   ✳ Divide by 3, then by 4

6 0.9 × 0.5 _____

   ✳ You know that 9 × 5 = 45

7 $\frac{2}{5}$ of 55 _____

   ✳ Find $\frac{1}{5}$ then multiply by 2

8 38 + 29 + 11 + 24 _____

   ✳ 29 + 11 first

9 7.1 × 5 _____

   ✳ The result must be about 35

10 17 × 7 _____

   ✳ The units digit of the result is 9

11 What temperature is 2 degrees lower than ⁻2°C?

   _____°C

12 Round 51 585 to 3 significant figures.

   _____

13 Write $\frac{5}{4}$ as a mixed number. _____

14 What proportion of the rectangle is blue?

   _____

15 Given that 4000 ÷ 50 = 80, what is 4000 ÷ 25?

   _____

16 What is the remainder when 80 is divided by 9?

   _____

17 By how much is 5 × 13 greater than 3 × 15?

   _____

18 Ivan buys 4 burgers priced at £1.95 each and 4 cans of cola priced at 65p each.

   How much change will he receive from a £20 note?

   £ _____

19 Oliver is thinking of a number and gives the following clues:

   "My number is:

   • between 15 and 45
   • a multiple of 6
   • a factor of 48"

   What number is Oliver thinking of?

   _____

20 On this tower of bricks, the number on a brick is the **product** of the numbers on the two bricks supporting it.

   What number is on the top brick? _____

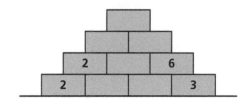

Test 16

For all of the questions in this test, do the calculation entirely in your head with no written 'working' and just write down the answer.

1   Round 5670 to the nearest 500 _____

2   What is 20% of £500?   £ _____

3   Two frogs, Hop and Skip, share 12 slugs so that Hop eats 2 more than Skip.

    How many slugs does Skip eat? _____

4   Blue has the four number cards below.

    What is the largest even number she can make by placing all four cards side by side? _____

5   When Lennie thought of a negative integer, added 2 and then multiplied by 3, the result was 3

    What number did Lennie think of? _____

6   If $a = 3$, $b = {}^-5$ and $c = 4$, what is the value of $ac + b$? _____

7   What is the next number in this sequence?

    100, 93, 86, 79, 72, _____

8   Sandy has a box of pencils. He decides that a quarter of them are too short to use and puts them in the bin. A third of the others need sharpening, but the remaining 8 are ready to use.

    How many pencils were in Sandy's box at the start? _____

9   Calculate the area of the rectangle.
    _____ cm²

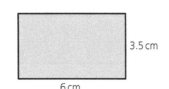

3.5 cm

6 cm

10 What is the reading on the scale? _____

0   1

11 Two angles of a triangle are 27° and 123°.

    What size is the third angle? _____°

12 What is the volume of a cuboid measuring 7 cm by 7 cm by 10 cm?

    _____ cm³

13 Honor and her group left at 9.45 p.m. for an overnight walk on Dartmoor and reached their destination at 7.05 a.m. How long did the journey take?

    _____ hours _____ minutes

Questions 14 to 16 refer to this list of the marks one pupil gained in 5 tests.

75   79   71   78   72

14 What is the range? _____

15 What is the median? _____

16 What is the mean? _____

17 The Carroll diagram shows some information about children in a class.

    How many more girls are there than boys in the class? _____ more

	Boys	Girls
Do not like sprouts	2	1
Like sprouts	7	10

18 What is the largest number less than 200 that is divisible exactly by 3? _____

19 The product of two integers is 42 and the difference between them is 11

    What is the sum of the integers? _____

20 Shelly has 15 hair bands. 6 are brown, a third of the others are white and the remainder are black.

    How many black hair bands does Shelly have? _____

Test 17

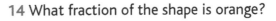

Final score

$$\frac{}{20}\ \_\_\_\ \%$$

For all of the questions in this test, do the calculation entirely in your head with no written 'working' and just write down the answer.

In questions 1 to 10 you are reminded of 10 useful strategies which may help you in later questions.

1 21 × 17 _____

    ✳ *Take easier steps when you multiply*

2 1284 ÷ 4 _____

    ✳ *Double or halve*

3 42 × 5 _____

    ✳ *Use your 10s and 2s*

4 124 − 59 _____

    ✳ *Take easier steps when you subtract*

5 350 ÷ 14 _____

    ✳ *Use factors to divide*

6 2.5 × 6 _____

    ✳ *Use known facts*

7 $\frac{4}{5}$ of 70 _____

    ✳ *Use a step-by-step approach*

8 3.9 + 1.5 + 11.1 + 2.4 _____

    ✳ *Group when you add*

9 5 × 39 _____

    ✳ *Approximate the result*

10 11 × 41 _____

    ✳ *Check using the units digit of the result*

11 What temperature is 5 degrees lower than 3°C?

_____ °C

12 Round 19.99 to 1 decimal place. _____

13 Write $\frac{9}{24}$ in its simplest form.

_____

14 What fraction of the shape is orange?

_____

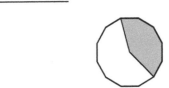

15 Given that 21 × 18 = 378, what is 22 × 18?

_____

16 What is the remainder when 90 is divided by 11?

_____

17 By how much is 10^2 greater than 9^2?

_____

18 Brodie buys 4 cans of cola priced at 75p each and 3 pork pies priced at £1.10 each.

How much change will she receive from a £20 note?

£ _____

19 Fatima is thinking of a 3-digit number and gives the following clues:

"My number is:

• less than 200
• a square number

and the digits add to 4"

What number is Fatima thinking of?

_____

20 On this tower of bricks, the number on each brick is the sum of the numbers on the two bricks supporting it.

What number is on the brick marked with a star (*)?

_____

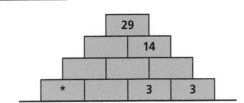

Test 18

For all of the questions in this test, do the calculation entirely in your head with no written 'working' and just write down the answer.

1 Round 3381 to the nearest 500 _____

2 What is 15% of £50? £ _____

3 Samantha and Natalie share 2 sliced pizzas in the ratio 3:5

   How many slices does Samantha get? _____

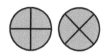

4 Lennie has the four number cards below.

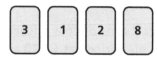

   What is the number closest to 100 he can make by placing some of the cards side by side? _____

5 When Lola thought of a number, multiplied it by 3 and then subtracted 2, the result was 1

   What number did Lola think of? _____

6 If $a = 1$ and $b = {}^-1$, what is the value of $a^2 + b^2$? _____

7 What is the next number in this sequence?

   1, 2, 6, 22, _____

8 Gina has a tin of foreign coins. She puts them in bags of 10 coins and she is just 1 coin short of filling a last bag. She sells the full bags of coins at 90p each and gives the money to charity. The charity receives £18

   How many coins were in Gina's tin? _____

9 Calculate the perimeter of the rectangle. _____ cm

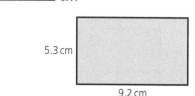

5.3 cm

9.2 cm

10 What is the reading on the scale? _____

```
 ⁻2        ⁻1        0        1
└┴┴┴┴┴┴┴┴┴┴┴┴┴┴┴┴┴┴┴┴┴┴┴┴┴┴┴┴┴┘
              ↑
```

11 Two angles of a triangle are 48° and 90°.

   What size is the third angle? _____°

12 A cube has volume 1000 cm³. What is the length of an edge of the cube?

   _____ cm

13 A $2\frac{3}{4}$ hour play ended at 21:55

   At what time did it start? _____ : _____

Questions 14 to 16 refer to this list of the marks one pupil gained in 5 tests.

13 17 12 14 19

14 What is the range? _____

15 What is the median? _____

16 What is the mean? _____

17 The Venn diagram shows some information about children at a party.

   How many people were not in fancy dress? _____

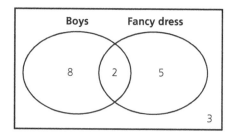

18 Which number is exactly half way between 44 and 72? _____

19 The sum of two integers (whole numbers) is 50 and the difference between them is 22

   What is the smaller number? _____

20 A factory has 160 workers. A quarter of the workers are men.

   How many women work in the factory?

   _____

Test 19

For all of the questions in this test, do the calculation entirely in your head with no written 'working' and just write down the answer.

In questions 1 to 10 you are reminded of 10 useful strategies which may help you in later questions.

1   31×39 _____

   ✳ 40×31 and then subtract 31

2   27×4 _____

   ✳ Double and then double again

3   $160 \div 5$ _____

   ✳ Divide by 10 then multiply by 2

4   $78 + 46$ _____

   ✳ Add 80 to 46 then subtract 2

5   $360 \div 24$ _____

   ✳ Divide by 6, then by 4

6   12×0.3 _____

   ✳ You know that $12 \times 3 = 36$

7   $\frac{3}{8}$ of 64 _____

   ✳ Find $\frac{1}{8}$ then multiply by 3

8   $47 + 34 + 15 + 33$ _____

   ✳ $47 + 33$ first

9   20.1×8 _____

   ✳ The result must be about 160

10 17×9 _____

   ✳ The units digit of the result is 3

11 How many degrees warmer is 2 °C than ⁻2 °C?

   _____ degrees

12 Round 107 546 to 4 significant figures.

   _____

13 Write $\frac{7}{3}$ as a mixed number. _____

14 What proportion of the rectangle is green?

   _____

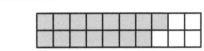

15 Given that $770 \div 35 = 22$, what is 22×70?

   _____

16 What is the remainder when 90 is divided by 7?

   _____

17 By how much is 7×14 greater than 4×17?

   _____

18 Beth buys 2 pizzas priced at £2.95 each and 4 bottles of juice priced at 75p each.

   How much change will she receive from a £10 note?

   £ _____

19 Bernard is thinking of a number and gives the following clues:

   "My number is:

   • between 100 and 200
   • a multiple of 12
   • 1 less than a multiple of 5"

   What number is Bernard thinking of?

   _____

20 On this tower of bricks, the number on a brick is the **product** of the numbers on the two bricks supporting it.

   What number is on the top brick? _____

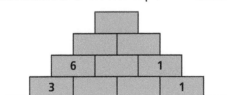

Test 20

For all of the questions in this test, do the calculation entirely in your head with no written 'working' and just write down the answer.

1 Round 34 070 to the nearest 100

_____

2 What is 15% of £200? £ _____

3 Two cats, Tabitha and Sally, share a packet of cat treats in the ratio 3:2

Sally eats 8 treats. How many treats does Tabitha eat? _____

4 Rani has the five number cards below.

What is the largest odd number she can make by placing all five cards side by side?

_____

5 When Lesley thought of a number, added 1 and then multiplied by 4, the result was 32

What number did Lesley think of? _____

6 If $a = 2$, $b = {}^-2$ and $c = {}^-5$, what is the value of $ab + ac$? _____

7 What is the next number in this sequence?

84, 78, 72, 66, 60, _____

8 Sandra has a tin of buttons. Half are white, a third are black and the other 6 are grey.

How many buttons are in Sandra's tin?

_____

9 Calculate the area of the rectangle.
   _____ cm²

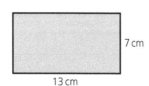

10 What is the reading on the scale? _____

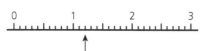

11 Two angles of a triangle are 28° and 102°.

What size is the third angle? _____°

12 What is the volume of a cuboid measuring 6 cm by 7 cm by 5 cm?

_____cm³

13 Elaine left home at 9.20 a.m. and walked to her friend's house, arriving at 10.05 a.m. After 30 minutes with her friend, Elaine walked home again at the same speed.

At what time did she arrive home?

_____ a.m./p.m.

Questions 14 to 16 refer to this list of the marks one pupil gained in 5 tests.

64 70 76 60 90

14 What is the range? _____

15 What is the median? _____

16 What is the mean? _____

17 The Carroll diagram shows some information about pupils in a class.

How many more pupils are aged 10 than are aged 11?

_____

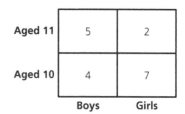

18 What is the largest number less than 100 that is divisible exactly by 13?

_____

19 The product of two integers is 60 and their sum is 19

What is the larger integer? _____

20 Margaux has 11 dolls. 3 have red dresses and a quarter of the others have blue dresses.

How many have blue dresses? _____

Test 21

For all of the questions in this test, do the calculation entirely in your head with no written 'working' and just write down the answer.

In questions 1 to 10 you are reminded of 10 useful strategies which may help you in later questions.

1 28 × 12 _____

  ✱ *Take easier steps when you multiply*

2 336 ÷ 8 _____

  ✱ *Double or halve*

3 48 × 5 _____

  ✱ *Use your 10s and 2s*

4 214 – 38 _____

  ✱ *Take easier steps when you subtract*

5 682 ÷ 22 _____

  ✱ *Use factors to divide*

6 3.9 × 4 _____

  ✱ *Use known facts*

7 $\frac{2}{3}$ of 123 _____

  ✱ *Use a step-by-step approach*

8 14.7 + 2.8 + 10.3 + 2.2 _____

  ✱ *Group when you add*

9 9 × 21 _____

  ✱ *Approximate the result*

10 12 × 13 _____

  ✱ *Check using the units digit of the result*

11 What temperature is 11 degrees lower than 6 °C?

  _____°C

12 Round 10.515 to 2 decimal places.

  _____

13 Write $\frac{8}{20}$ in its simplest form. _____

14 Write the purple parts of these two shapes as a mixed fraction.

  _____

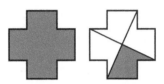

15 Given that 37 × 47 = 1739, what is 37 × 48?

  _____

16 What is the remainder when 60 is divided by 13?

  _____

17 By how much is 11² greater than 10²?

  _____

18 Tanya buys 2 mugs of coffee and 2 muffins. The total cost is exactly £5

  The muffins are priced at £1.35 each.

  What is the price of 1 mug of coffee?

  £ _____

19 Faisal is thinking of a 2-digit number and gives the following clues:

  "My number is:

  • a multiple of 11
  • even
  • 1 less than a multiple of 5"

  What number is Faisal thinking of?

  _____

20 On this tower of bricks, the number on each brick is the sum of the numbers on the two bricks supporting it.

  What number is on the brick marked with a star (*)? _____

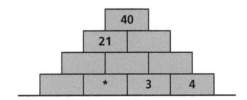

Test 22

For all of the questions in this test, do the calculation entirely in your head with no written 'working' and just write down the answer.

1   Round 76 849 to the nearest 100 _____

2   What is 35% of £200? £ _____

3   Eve, Adam and Julia share 20 chocolates in the ratio 3 : 3 : 4

How many chocolates does Adam get? _____

4   Jay has the seven number cards below.

What is the number closest to 1 million he can make by placing cards side by side? _____

5   When Aimee thought of a number, multiplied it by 2 and then subtracted 5, the result was ⁻1

What number did Aimee think of? _____

6   If $a = 2$ and $b = ^-3$, what is the value of $a^2 + b$? _____

7   What is the next number in this sequence?

1, 4, 19, 94, _____

8   Hollie has a collection of seashells. She puts them in bags of 5 shells and sells the bags of shells at 80p each. She receives a total of £9.60

How many shells did Hollie have in her collection? _____

9   Calculate the perimeter of the rectangle. _____ cm

8.9 cm

9.2 cm

10   What is the reading on the scale? _____

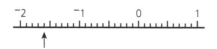

11   Two angles of a triangle are 47° and 76°.

What size is the third angle? _____ °

12   A cuboid has volume 200 cm³. It is 10 cm long and 5 cm wide.

How tall is the cuboid? _____ cm

13   A 1 hour 50 minutes concert ended at 8.25 p.m.

At what time did it start? _____ p.m.

Questions 14 to 16 refer to this list showing the lengths in centimetres of 5 strips of wood.

29   33   21   20   17

14   What is the range? _____ cm

15   What is the median? _____ cm

16   What is the mean? _____ cm

17   The Venn diagram shows some information about people watching a rugby match.

How many people watched the match? _____

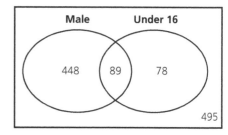

18   Which number is exactly half way between 100 and 122? _____

19   The sum of two integers (whole numbers) is 40 and the difference between them is 20

What is the product of the numbers? _____

20   In this pictogram, one symbol represents **two** beetles.

Type A	🐞 🐞 🐞 🐞
Type B	🐞 🐞 🐞
Type C	🐞 🐞 🐞 🐞 🐞

What was the total number of beetles found? _____

Answers

Test 1

1　87
2　£3.10
3　240
4　16
5　34
6　4.2
7　18
8　78
9　7.8
10　184
11　$^{-}3\,°C$
12　5.5
13　$\frac{3}{4}$
14　$\frac{7}{12}$
15　1551
16　2
17　10
18　£2.21
19　14
20　24

Test 2

1　4700
2　8 kg
3　4
4　23 478
5　3
6　23
7　21
8　25
9　32 cm
10　1.8
11　10°
12　60 cm³
13　21:15
14　10
15　9
16　8
17　5
18　66
19　17
20　6

Test 3

1　76
2　£33.60
3　26
4　111
5　28
6　4.0 (or 4)
7　36
8　97
9　40.8
10　102

11　2 °C
12　40
13　$1\frac{1}{4}$
14　$\frac{5}{8}$ or 62.5%
15　246
16　4
17　10
18　£1.57
19　24
20　24

Test 4

1　41 000
2　24 ml
3　80p
4　987
5　2
6　9
7　16
8　15
9　24 cm²
10　2.4
11　70°
12　160 cm³
13　2 hours 15 minutes
14　6
15　7
16　3
17　35%
18　102
19　32
20　12

Test 5

1　209
2　16
3　615
4　47
5　22
6　5.2
7　48
8　9.5
9　95
10　117
11　$^{-}4\,°C$
12　3.49
13　$\frac{2}{3}$
14　$\frac{1}{2}$
15　2108
16　2
17　7
18　£1.40
19　15
20　1

Test 6

1　410 000
2　£1.50
3　30
4　1352
5　7
6　15
7　63
8　88
9　22 cm
10　$^{-}0.35$
11　20°
12　32 cm³
13　18:40
14　8
15　8
16　7.5
17　35
18　84
19　12
20　20

Test 7

1　319
2　£17.60
3　44
4　204
5　13
6　5.6
7　28
8　120
9　19.6
10　70
11　3 °C
12　40
13　$\frac{10}{3}$
14　$\frac{2}{5}$ or 40%
15　21
16　5
17　50
18　£6.60
19　42
20　48

Test 8

1　50 000
2　60 kg
3　£1.50
4　496
5　3
6　6
7　122
8　14
9　10 cm²
10　2.2

11 28°
12 70 cm³
13 2 hours 40 minutes
14 5
15 9
16 8
17 75%
18 92
19 60
20 £8.50

Test 9

1 231
2 28
3 240
4 64
5 33
6 5.6
7 28
8 14.9
9 114
10 156
11 ⁻1°C
12 47.5
13 $\frac{2}{3}$
14 70%
15 2322
16 1
17 9
18 £6.10
19 48
20 1

Test 10

1 12 300
2 £5.40
3 4
4 198
5 5
6 5
7 171
8 123
9 28 cm
10 0.48
11 74°
12 60 cm³
13 18:55
14 8
15 7.5
16 7.3
17 24
18 102
19 29
20 22

Test 11

1 1281
2 £13.80
3 88
4 146

5 15
6 10.8
7 57
8 89
9 8.2
10 144
11 ⁻5°C
12 55
13 $2\frac{1}{3}$
14 $\frac{4}{5}$ or 80%
15 24
16 6
17 60
18 £1.70
19 10
20 128

Test 12

1 100 000
2 2 litres
3 £8
4 87 530
5 4
6 8
7 32
8 26
9 34 cm²
10 0.18
11 50°
12 320 cm³
13 1 hour 20 minutes
14 9
15 17
16 16
17 $\frac{1}{2}$
18 98
19 36
20 16

Test 13

1 247
2 28
3 190
4 34
5 18
6 7 (or 7.0)
7 9
8 10
9 116
10 169
11 ⁻6°C
12 105.9
13 $\frac{5}{9}$
14 $\frac{3}{7}$
15 672
16 2
17 15

18 £4
19 64
20 7

Test 14

1 300
2 3 litres
3 5 more
4 42
5 7
6 3
7 41
8 623
9 29 cm
10 1.2
11 102°
12 500 cm³
13 19:55
14 18
15 18
16 17
17 140
18 57
19 26
20 66

Test 15

1 902
2 136
3 48
4 121
5 13
6 0.45
7 22
8 102
9 35.5
10 119
11 ⁻4°C
12 51 600
13 $1\frac{1}{4}$
14 $\frac{5}{8}$ or 62.5%
15 160
16 8
17 20
18 £9.60
19 24
20 48

Test 16

1 5500
2 £100
3 5
4 8740
5 ⁻1
6 7
7 65
8 16
9 21 cm²
10 0.7

11 30°
12 490 cm³
13 9 hours 20 minutes
14 8
15 75
16 75
17 2 more
18 198
19 17
20 6

Test 17

1 357
2 321
3 210
4 65
5 25
6 15
7 56
8 18.9
9 195
10 451
11 ⁻2 °C
12 20.0 ('tenths' 0 essential)
13 $\frac{3}{8}$
14 $\frac{5}{12}$
15 396
16 2
17 19
18 £13.70
19 121
20 2

Test 18

1 3500
2 £7.50
3 3
4 83
5 1
6 2
7 86
8 209
9 29 cm
10 ⁻0.5
11 42°
12 10 cm
13 19:10
14 7
15 14
16 15
17 11
18 58
19 14
20 120

Test 19

1 1209
2 108
3 32

4 124
5 15
6 3.6
7 24
8 129
9 160.8
10 153
11 4 degrees
12 107 500
13 $2\frac{1}{3}$
14 $\frac{3}{4}$ or 75%
15 1540
16 6
17 30
18 £1.10
19 144
20 24

Test 20

1 34 100
2 £30
3 12
4 76 403
5 7
6 ⁻14
7 54
8 36
9 91 cm²
10 1.2
11 50°
12 210 cm³
13 11.20 a.m.
14 30
15 70
16 72
17 4
18 91
19 15
20 2

Test 21

1 336
2 42
3 240
4 176
5 31
6 15.6
7 82
8 30
9 189
10 156
11 ⁻5 °C
12 10.52
13 $\frac{2}{5}$
14 $1\frac{1}{4}$
15 1776
16 8

17 21
18 £1.15
19 44
20 9

Test 22

1 76 800
2 £70
3 6
4 1 002 368
5 2
6 1
7 469
8 60
9 36.2 cm
10 ⁻1.6
11 57°
12 4 cm
13 6.35 p.m.
14 16 cm
15 21 cm
16 24 cm
17 1110
18 111
19 300
20 23

Test 23

1 748
2 144
3 66
4 146
5 18
6 3 (or 3.0)
7 36
8 208
9 25.8
10 112
11 10 degrees
12 408 000
13 $2\frac{2}{5}$
14 $\frac{5}{8}$ or 62.5%
15 216
16 6
17 70
18 £11.15
19 80
20 1

Test 24

1 29 150
2 £110
3 14
4 36
5 5
6 2
7 33
8 18
9 95 cm²

10 4.1
11 59°
12 1320 cm³
13 12.05 p.m.
14 8 cm
15 123 cm
16 124 cm
17 $\frac{1}{5}$
18 108
19 14
20 $\frac{2}{3}$

Test 25
1 472
2 26
3 360
4 995
5 38
6 6.5
7 150
8 50
9 232
10 182
11 ⁻2°C
12 1.3
13 $\frac{1}{4}$
14 40%
15 1276
16 3
17 80
18 £2.10
19 49
20 2

Test 26
1 305000
2 12 kg
3 £40
4 1023
5 20
6 7
7 161
8 42
9 19 m
10 ⁻1.3
11 108°
12 5 cm
13 6.35 p.m.
14 2.2 kg
15 3.3 kg
16 3 kg
17 59%
18 80
19 17
20 32

Test 27
1 2142
2 144

3 28
4 186
5 14
6 0.96
7 68
8 156
9 39.8
10 96
11 11 degrees
12 305000
13 $2\frac{1}{4}$
14 $\frac{1}{4}$ or 25%
15 46
16 2
17 40
18 £3.70
19 36
20 5

Test 28
1 176000
2 £450
3 27
4 20
5 8
6 0
7 88
8 270
9 135 cm²
10 0.3
11 64°
12 210 cm³
13 43 minutes
14 64
15 64
16 65
17 $\frac{2}{3}$
18 48
19 7
20 65

Test 29
1 429
2 23
3 370
4 86
5 13
6 44
7 165
8 72.7
9 261
10 90
11 ⁻5°C
12 0.02
13 $\frac{2}{5}$
14 $\frac{1}{6}$ more
15 696

16 5
17 80
18 £0.80
19 81
20 6

Test 30
1 4800
2 £9
3 £5
4 8610
5 8
6 11
7 341
8 100
9 12.6 cm
10 ⁻0.2
11 39°
12 6 cm²
13 18:55
14 £19.00
15 £19.50
16 £20.00
17 40%
18 42
19 16
20 £56

Test 31
1 1271
2 180
3 48
4 222
5 16
6 0.81
7 £2.24
8 140
9 59.8
10 299
11 6 degrees
12 408000
13 $3\frac{2}{3}$
14 $\frac{2}{3}$ or 66.$\dot{6}$%
15 32
16 1
17 30
18 £1.60
19 16
20 1

Test 32
1 305000
2 £3.60
3 46
4 476
5 7
6 ⁻7
7 16
8 18

Mathematics Workbook: Mental Arithmetic Age 9–11 published by Galore Park

9 8 cm²
10 0.47
11 81°
12 100 cm³
13 33 minutes
14 14
15 39
16 40
17 35%
18 63
19 16
20 71

Test 33

1 399
2 220
3 29
4 175
5 13
6 3.6
7 8
8 238
9 44.4
10 156
11 17 degrees
12 16.5
13 $1\frac{1}{5}$

14 $\frac{4}{7}$

15 4361
16 4
17 30
18 2
19 9
20 4

Test 34

1 31 000
2 £75
3 30 cl
4 300
5 6
6 20
7 83
8 4 dozen
9 9 cm
10 ⁻0.8
11 140°
12 3 cm
13 8.55 p.m.
14 0
15 2
16 2.5
17 30%
18 1250
19 8
20 13

Test 35

1 187

2 152
3 17
4 116
5 23
6 10.8
7 25
8 199
9 195
10 112
11 11 degrees
12 390 600
13 $2\frac{1}{5}$

14 $\frac{2}{5}$ or 40%

15 1440
16 2
17 20
18 4
19 66
20 1

Test 36

1 2500
2 £10
3 3:2
4 84
5 4
6 4
7 171
8 8
9 100 cm²
10 12.3
11 130°
12 56 cm³
13 15:50
14 0.6 seconds
15 15.0 seconds
16 15.0 seconds
17 55%
18 63
19 16
20 $\frac{2}{3}$

Test 37

1 418
2 176
3 46
4 285
5 18
6 5.6
7 26
8 218
9 240.6
10 144
11 6.5 degrees
12 16.5
13 $4\frac{1}{2}$

14 $\frac{1}{3}$ or 33.$\dot{3}$%

15 1739
16 1
17 19
18 £3.35
19 48
20 7

Test 38

1 235 000
2 £6
3 3:2
4 8 (8643)
5 4
6 8
7 45
8 2
9 20 cm
10 ⁻0.4
11 120°
12 6 cm
13 10.05 p.m.
14 105
15 35
16 44
17 22
18 49
19 13
20 24

Test 39

1 495
2 99
3 170
4 135
5 23
6 13.2
7 69
8 87
9 99.9
10 112
11 17 degrees
12 4000
13 $2\frac{2}{5}$

14 $\frac{4}{7}$

15 2688
16 2
17 10
18 £2
19 53
20 6

Test 40

1 44 500
2 £150
3 8
4 4876
5 16
6 10
7 128

8 44%
9 8 cm²
10 22
11 140°
12 500 cm³
13 11.30 a.m.
14 0.5 seconds
15 32.4 seconds
16 32.3 seconds
17 2
18 38
19 15
20 10

Test 41

1 441
2 492
3 44
4 303
5 17
6 0.6
7 £1.40
8 150
9 £119.60
10 182
11 5 degrees
12 300
13 $1\frac{3}{4}$
14 $\frac{1}{4}$ or 25%
15 576
16 3
17 20
18 £1.60
19 73
20 2

Test 42

1 78 000
2 £4.50
3 34
4 915
5 11
6 3
7 86
8 25
9 22.5 cm²
10 0.24
11 35°
12 4 cm
13 15 minutes
14 32
15 29
16 27.5
17 6000
18 72
19 9
20 84

Test 43

1 357

2 192
3 59
4 402
5 19
6 5.6
7 162
8 11.93
9 60.4
10 182
11 5 degrees
12 17
13 $1\frac{3}{5}$
14 $\frac{7}{12}$ or 58.$\dot{3}$%
15 6804
16 2
17 20
18 3
19 28
20 7

Test 44

1 410 000
2 £1.20
3 36
4 3300
5 3
6 7
7 28
8 32
9 20 cm
10 ⁻1.6
11 106°
12 2.5 cm
13 10.10 p.m.
14 2
15 2
16 2.2
17 69
18 64
19 11
20 19

Test 45

1 253
2 176
3 800
4 163
5 16
6 13.2
7 £75
8 160
9 189
10 84
11 5 degrees
12 38 500
13 $3\frac{2}{5}$
14 $\frac{1}{4}$ or 25%
15 221
16 1

17 40
18 3
19 99
20 4

Test 46

1 75 500
2 £13.20
3 4:3
4 1013
5 10
6 13
7 94
8 24
9 14 cm²
10 5.27
11 70°
12 60 cm³
13 15:15
14 1.8 seconds
15 71.2 seconds
16 71.0 seconds
17 60%
18 80
19 14
20 11

Test 47

1 378
2 292
3 94
4 355
5 26
6 0.3
7 160
8 £11.55
9 149.4
10 256
11 4.5 degrees
12 0.05
13 $3\frac{1}{4}$
14 $\frac{5}{12}$ or 41.$\dot{6}$%
15 27
16 1
17 50
18 3
19 60
20 1

Test 48

1 461 000
2 £0.24
3 21
4 8
5 $\frac{2}{3}$
6 ⁻18
7 7.5
8 £2.40
9 43 cm

Mathematics Workbook: Mental Arithmetic Age 9–11 published by Galore Park

10 ⁻3.6
11 57°
12 6 cm
13 7.30 p.m.
14 186
15 100
16 110
17 $\frac{1}{3}$
18 55
19 $\frac{1}{4}$
20 £16.20

Test 49

1 638
2 54
3 8.8
4 5177
5 37

6 0.5
7 £20.40
8 380
9 £239.60
10 299
11 18 degrees
12 35 000
13 $\frac{23}{5}$
14 $\frac{7}{20}$ or 35%
15 6004
16 4
17 60
18 £11.20
19 43
20 8

Test 50

1 145 000

2 £225
3 44
4 49 865
5 $\frac{3}{4}$
6 9
7 41
8 9
9 11.2 cm²
10 47.2
11 47°
12 34 cm²
13 11.58 a.m.
14 28
15 33
16 30
17 40
18 1002
19 20
20 22

Test 23

For all of the questions in this test, do the calculation entirely in your head with no written 'working' and just write down the answer.

In questions 1 to 10 you are reminded of 10 useful strategies which may help you in later questions.

1 68 × 11 _____

　✳ *70 × 11 or 68 × 10 first*

2 36 × 4 _____

　✳ *Double and then double again*

3 330 ÷ 5 _____

　✳ *Divide by 10 then multiply by 2*

4 89 + 57 _____

　✳ *Add 90 to 57 then subtract 1*

5 576 ÷ 32 _____

　✳ *Divide by 4, then by 8*

6 15 × 0.2 _____

　✳ *You know that 15 × 2 = 30*

7 $\frac{4}{5}$ of 45 _____

　✳ *Find $\frac{1}{5}$ then multiply by 4*

8 95 + 48 + 5 + 60 _____

　✳ *95 + 5 first*

9 12.9 × 2 _____

　✳ *The result must be about 26*

10 14 × 8 _____

　✳ *The units digit of the result is 2*

11 How many degrees warmer is 7 °C than ⁻3 °C?

_____ degrees

12 Round 408 463 to 3 significant figures.

_____

13 Write $\frac{12}{5}$ as a mixed number. _____

14 What proportion of the rectangle is blue?

_____

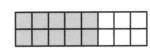

15 Given that 864 ÷ 36 = 24, what is 18 × 12?

_____

16 What is the remainder when 300 is divided by 7?

_____

17 By how much is 9 × 12 greater than 2 × 19?

_____

18 Georgie buys 2 DVDs priced at £10.95 each and a CD priced at £6.95

How much change will she receive from two £20 notes?

£ _____

19 José is thinking of a number and gives the following clues:

"My number is:

· between 50 and 100
· a multiple of 4
· 1 less than a square number."

What number is José thinking of? _____

20 On this tower of bricks, the number on a brick is the **product** of the numbers on the two bricks supporting it.

What number is on the brick marked with a star (*)?

_____

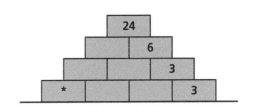

Test 24

For all of the questions in this test, do the calculation entirely in your head with no written 'working' and just write down the answer.

1 Round 29 172 to the nearest 50

_____

2 What is 11% of £1000? £ _____

3 Two dogs, Sam and Rosie, share a box of dog biscuits in the ratio 4:3

Sam eats 8 biscuits. How many biscuits were in the box? _____

4 Peter has the four number cards below.

196 is a 3-digit square number (14^2).

What is the largest 2-digit square number he can make by placing two of the cards side by side? _____

5 When Lydia thought of a number, subtracted 2 and then multiplied by 3, the result was 9

What number did Lydia think of? _____

6 If $a = 3$, $b = ^-1$ and $c = ^-2$, what is the value of $a + c - b$? _____

7 What is the next number in this sequence?

103, 89, 75, 61, 47, _____

8 Sean has a collection of 37 medals. 13 are war medals and a quarter of the rest are sports medals. The remainder are coronation medals.

How many coronation medals does Sean have? _____

9 Calculate the area of the rectangle.
_____ cm²

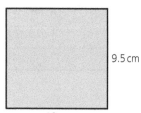

9.5 cm

10 cm

10 What is the reading on the scale? _____

0   2   4   6

11 Two angles of a triangle are 46° and 75°.

What size is the third angle? _____°

12 What is the volume of a cuboid measuring 10 cm by 11 cm by 12 cm? _____ cm³

13 Evelyn left home at 11.55 a.m. and walked to the post box. After posting the letter, she walked home again at the same speed, arriving at 12.15 p.m.

At what time did Evelyn post the letter? _____ a.m. /p.m.

Questions 14 to 16 refer to this list showing the heights (in cm) of Toni's high jumps.

123 125 122 129 121

14 What is the range? _____ cm

15 What is the median? _____ cm

16 What is the mean? _____ cm

17 The Carroll diagram shows some information about friends at a party.

What fraction of the girls do not like trifle?

_____

	Boys	Girls
Do not like trifle	3	1
Like trifle	5	4

18 What is the smallest number greater than 100 that is divisible exactly by 12? _____

19 The product of two integers, both less than 10, is 45

What is the sum of the integers? _____

20 Nigel has 15 model vehicles. 3 are buses, 2 are vans and the rest are cars.

What fraction of his model vehicles is cars?

_____

Test 25

For all of the questions in this test, do the calculation entirely in your head with no written 'working' and just write down the answer.

In questions 1 to 10 you are reminded of 10 useful strategies which may help you in later questions.

1 59×8 _____

   * *Take easier steps when you multiply*

2 $208 \div 8$ _____

   * *Double or halve*

3 72×5 _____

   * *Use your 10s and 2s*

4 $1032 - 37$ _____

   * *Take easier steps when you subtract*

5 $456 \div 12$ _____

   * *Use factors to divide*

6 1.3×5 _____

   * *Use known facts*

7 $\frac{3}{8}$ of 400 _____

   * *Use a step-by-step approach*

8 $18.2 + 9.3 + 10.7 + 11.8$ _____

   * *Group when you add*

9 8×29 _____

   * *Approximate the result*

10 14×13 _____

   * *Check using the units digit of the result*

11 What temperature is 4 degrees higher than $^-6\,°C$?

_____°C

12 Round 1.345 to 1 decimal place. _____

13 Write $\frac{6}{24}$ in its simplest form.

_____

14 What percentage of this shape is blue?

_____%

15 Given that $48 \times 22 = 1056$, what is 58×22?

_____

16 What is the remainder when 80 is divided by 7?

_____

17 By how much is 12^2 greater than 8^2?

_____

18 Trevor buys 4 yoghurts on special offer.

Buy two, get a third free

The normal price of 1 yoghurt is 70p.

How much does Trevor pay? £ _____

19 Phil is thinking of a 2-digit number and gives the following clues:

"My number is:

• a multiple of 7
• odd
• 1 less than a multiple of 10"

What number is Phil thinking of? _____

20 On this tower of bricks, the number on each brick is the sum of the numbers on the two bricks supporting it.

What number is on the brick marked with a star (*)?

_____

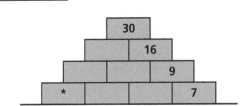

31

Test 26

For all of the questions in this test, do the calculation entirely in your head with no written 'working' and just write down the answer.

1 Round 305 690 to the nearest 5000

_____

2 What is 15% of 80 kg? _____ kg

3 Tom and Dixie share a £100 prize in the ratio of their ages. Tom is aged 8 and Dixie is 4 years older than Tom.

How much will Tom receive?   £ _____

4 Joanna has the six number cards below.

What is the number closest to 1000 she can make by placing cards side by side?

_____

5 When Donald thought of a number, multiplied it by 3 and then subtracted 2, the result was 58

What number did Donald think of?

_____

6 If $a = 6$ and $b = {}^-5$, what is the value of $2a + b$?

_____

7 What is the next number in this sequence?

1, 5, 17, 53, _____

8 Sophie has a collection of elephants. Half of them are china, a third of them are wood and the other 7 are metal.

How many elephants does Sophie have in her collection? _____

9 Calculate the perimeter of the rectangle.
_____ m

0.8 m
8.7 m

10 What is the reading on the scale? _____

$^-4$ $^-2$ 0   2

11 Two angles of a triangle are 36° and 36°.

What size is the third angle? _____°

12 A cuboid has volume 120 cm³. It is 12 cm long and 2 cm wide.

How tall is the cuboid? _____ cm

13 A 1 hour 50 minutes concert ended at 8.25 p.m.

At what time did it start? _____ p.m.

Questions 14 to 16 refer to this list showing the masses in kilograms of 5 large stones.

3.7 3.3 2.2 4.0 1.8

14 What is the range? _____ kg

15 What is the median? _____ kg

16 What is the mean? _____ kg

17 The Venn diagram shows some information about people at a concert.

What percentage of the people were under 16?

_____ %

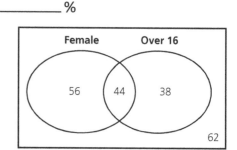

18 Which number is exactly half way between 40 and 120? _____

19 The sum of two integers (whole numbers) is 30 and the difference between them is 4

What is the larger number? _____

20 In this pictogram, showing the sales of drinks, one symbol 🍵 represents **four** drinks.

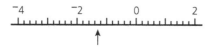

Tea	🍵 🍵 🍵 🍵 🍵 🍵
Coffee	🍵 🍵 🍵

What is the total number of drinks sold?

_____

Test 27

For all of the questions in this test, do the calculation entirely in your head with no written 'working' and just write down the answer.

In questions 1 to 10 you are reminded of 10 useful strategies which may help you in later questions.

1 51 × 42 _____

* Take easier steps when you multiply

2 36 × 4 _____

* Double or halve

3 140 ÷ 5 _____

* Use your 10s and 2s

4 119 + 67 _____

* Take easier steps when you add

5 210 ÷ 15 _____

* Use factors to divide

6 1.2 × 0.8 _____

* Use known facts

7 $\frac{4}{5}$ of 85 _____

* Use a step-by-step approach

8 29 + 16 + 61 + 50 _____

* Group when you add

9 19.9 × 2 _____

* Approximate the result

10 16 × 6 _____

* Check using the units digit of the result

11 How many degrees cooler is ⁻7 °C than 4 °C?

_____ degrees

12 Round 304 509 to 3 significant figures.

_____

13 Write $\frac{9}{4}$ as a mixed number. _____

14 What proportion of the square is purple?

_____

15 Given that 1288 ÷ 23 = 56, what is 1288 ÷ 28?

_____

16 What is the remainder when 50 is divided by 6?

_____

17 By how much is 6 × 12 greater than 2 × 16?

_____

18 Brenda buys 3 muffins priced at £1.15 each and 3 cups of tea priced at 95p each.

How much change will she receive from a £10 note?

£ _____

19 Anatoly is thinking of a number and gives the following clues:

"My number is:

• between 10 and 50
• a factor of 72
• a square number."

What number is Anatoly thinking of?

_____

20 On this tower of bricks, the number on a brick is the **product** of the numbers on the two bricks supporting it.

What number is on the brick marked with a star (*)? _____

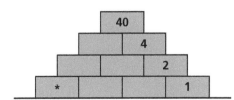

Test 28

For all of the questions in this test, do the calculation entirely in your head with no written 'working' and just write down the answer.

1 Round 175 840 to the nearest 1000
_____

2 What is 90% of £500?   £ _____

3 There are between 25 and 30 toffees in a bag. Thomasina and Jeremiah eat all the toffees, sharing them in the ratio 5 : 4

How many toffees were in the bag?
_____

4 Ralf has the three number cards below.

What is the difference between the largest and smallest 2-digit numbers he can make by placing cards side by side? _____

5 When Linda thought of a number, added 2 and then multiplied by 4, the result was 40

What number did Linda think of? _____

6 If $a = 4$ and $b = {}^-2$, what is the value of $a + 2b$?
_____

7 What is the next number in this sequence?

123, 116, 109, 102, 95, _____

8 Stuart has an album of stamps. A third are British and the rest are foreign. He has 180 foreign stamps.

What is the total number stamps in Stuart's album? _____

9 Calculate the area of the rectangle.
_____ cm²

4.5 cm

30 cm

10 What is the reading on the scale? _____

0   1

11 Two angles of a triangle are 88° and 28°.

What size is the third angle? _____°

12 What is the volume of a cuboid measuring 5 cm by 6 cm by 7 cm?

_____ cm³

13 Elizabeth left home at 11.35 a.m. to visit her aunt, arriving at her aunt's house at 12.18 p.m.

How long did the journey take her?

_____ minutes

Questions 14 to 16 refer to this list of the marks one pupil gained in 5 tests.

66 64 69 62 64

14 What is the mode? _____

15 What is the median? _____

16 What is the mean? _____

17 The Carroll diagram shows some information about pupils in a class.

What fraction of the boys has blue eyes?

_____

Not blue eyes	3	4
Blue eyes	6	4
	Boys	Girls

18 What is the largest number less than 50 that is a multiple of both 4 and 6?

_____

19 The product of two integers is 42 and their sum is 13

What is the larger integer? _____

20 A farmer keeps pigs, sheep and cows. He has 132 animals. 12 of these are pigs and he has 10 more sheep than cows.

How many sheep does he have? _____

Test 29

For all of the questions in this test, do the calculation entirely in your head with no written 'working' and just write down the answer.

In questions 1 to 10 you are reminded of 10 useful strategies which may help you in later questions.

1   39×11 _____

* Take easier steps when you multiply

2   $184 \div 8$ _____

* Double or halve

3   74×5 _____

* Use your 10s and 2s

4   $145 - 59$ _____

* Take easier steps when you subtract

5   $208 \div 16$ _____

* Use factors to divide

6   5.5×8 _____

* Use known facts

7   $\frac{3}{4}$ of 220 _____

* Use a step-by-step approach

8   $35.5 + 14.7 + 22.5$ _____

* Group when you add

9   9×29 _____

* Approximate the result

10 15×6 _____

* Check using the units digit of the result

11 What temperature is 7 degrees lower than 2 °C?

_____°C

12 Round 0.017 to 2 decimal places. _____

13 Write $\frac{14}{35}$ in its simplest form.

_____

14 How much more of shape A is orange than shape B? Give your answer as a fraction of a shape.

_____ more

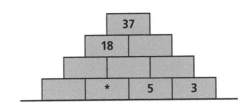

15 Given that $28 \times 24 = 672$, what is 29×24?

_____

16 What is the remainder when 40 is divided by 7?

_____

17 By how much is 9^2 greater than 1^2?

_____

18 Salome buys 2 cups of tea and 2 cakes. The total cost is exactly £4.50

The cakes are priced at £1.45 each. What is the price of 1 cup of tea?

£ _____

19 Francis is thinking of a 2-digit number and gives the following clues:

"My number is:
• a multiple of 9
• odd
• 1 more than a multiple of 10"

What number is Francis thinking of?

_____

20 On this tower of bricks, the number on each brick is the sum of the numbers on the two bricks supporting it.

What number is on the brick marked with a star (*)?

_____

Test 30

For all of the questions in this test, do the calculation entirely in your head with no written 'working' and just write down the answer.

1 Round 4824 to the nearest 50 _____

2 What is 45% of £20? £ _____

3 Sergei and Ahmed share £45 in the ratio 4:5

   How much more than Sergei does Ahmed get?
   £ _____

4 Jenny has the four number cards below.

   What is the largest multiple of 6 she can make by placing cards side by side? _____

5 When Martine thought of a number, multiplied it by 3 and then subtracted 2, the result was 22

   What number did Martine think of?
   _____

6 If $a = 5$ and $b = {}^-4$, what is the value of $3a + b$?
   _____

7 What is the next number in this sequence?

   1, 5, 21, 85, _____

8 Ramona has picked some daffodils in her garden. She bundles them in fives and sells the bundles of daffodils at 75p each for church funds. She takes a total of £15.00

   How many daffodils did Ramona pick?
   _____

9 Calculate the perimeter of the equilateral triangle. _____ cm

4.2 cm

10 What is the reading on the scale? _____

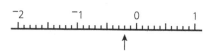

11 Two angles of a triangle are 74° and 67°.

   What size is the third angle? _____°

12 A cube has edges of length 1 cm. What is the total surface area of the cube?

   _____ cm²

13 A film lasting 2 hours 10 minutes ended at 21:05

   At what time did it start? _____:_____

Questions 14 to 16 refer to this list showing the prices in pounds of an iron at 5 different shops.

£19.00 £20.50 £19.50 £19.00 £22.00

14 What is the mode? £ _____

15 What is the median? £ _____

16 What is the mean? £ _____

17 The Venn diagram shows some information about people on a bus.

   What percentage of the females was under 16? _____%

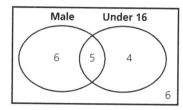

18 Which number is exactly half way between 20 and 64? _____

19 The sum of two integers (whole numbers) is 25 and the difference between them is 7

   What is the larger number? _____

20 In this pictogram, one symbol 🧸 represents one bear sold.

| Brown bears £5 each | 🧸 🧸 🧸 🧸 🧸 🧸 |
| Yellow bears £6.50 each | 🧸 🧸 🧸 🧸 |

   How much money was taken from the sale of the bears? £ _____

Test 31

For all of the questions in this test, do the calculation entirely in your head with no written 'working' and just write down the answer.

In questions 1 to 10 you are reminded of 10 useful strategies which may help you in later questions.

1 41 × 31 _____

   ✳ *Take easier steps when you multiply*

2 45 × 4 _____

   ✳ *Double or halve*

3 240 ÷ 5 _____

   ✳ *Use your 10s and 2s*

4 138 + 84 _____

   ✳ *Take easier steps when you add*

5 288 ÷ 18 _____

   ✳ *Use factors to divide*

6 0.9 × 0.9 _____

   ✳ *Use known facts*

7 $\frac{2}{5}$ of £5.60 £ _____

   ✳ *Use a step-by-step approach*

8 47 + 14 + 53 + 26 _____

   ✳ *Group when you add*

9 29.9 × 2 _____

   ✳ *Approximate the result*

10 13 × 23 _____

   ✳ *Check using the units digit of the result*

11 How many degrees cooler is ⁻5 °C than 1 °C?

   _____ degrees

12 Round 408 499 to 3 significant figures.

   _____

13 Write $\frac{11}{3}$ as a mixed number. _____

14 The diagram below shows a garden.

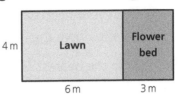

   What proportion of the garden is lawn?

   _____

15 Given that 1152 ÷ 18 = 64, what is 1152 ÷ 36?

   _____

16 What is the remainder when 40 is divided by 13?

   _____

17 By how much is 6 × 13 greater than 3 × 16?

   _____

18 Tilly buys 4 muffins priced at £1.12 each and 4 bottles of cola priced at 98p each.

   How much change will she receive from a £10 note?

   £ _____

19 Kai is thinking of a number and gives the following clues:

   "My number is:

   • between 15 and 45
   • a factor of 80
   • a square number."

   What number is Kai thinking of? _____

20 On this tower of bricks, the number on a brick is the **product** of the numbers on the two bricks supporting it.

   What number is on the brick marked with a star (*)?

   _____

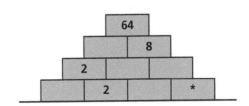

Test 32

For all of the questions in this test, do the calculation entirely in your head with no written 'working' and just write down the answer.

1 Round 305 999 to the nearest 5000

_____

2 What is 20% of £18? £ _____

3 There are between 41 and 50 sweets in a packet. When 6 friends share them fairly, there are 4 sweets left over.

How many sweets were in the packet?

_____

4 Robin has the four number cards below.

What is the number nearest to 500 he can make by placing cards side by side?

_____

5 When Louisa thought of a number, added 3 and then multiplied by 2, the result was 20

What number did Louisa think of? _____

6 If $a = ^-4$, $b = ^-4$ and $c = 5$, what is the value of $a + 2b + c$? _____

7 What is the next number in this sequence?

1, 2, 4, 7, 11, _____

8 Alicia has a collection of 48 souvenir leather bookmarks. Half are British and three-quarters of the others are European.

How many European bookmarks does Alicia have? _____

9 Calculate the area of the triangle.
_____ cm²

4 cm

4 cm

10 What is the reading on the scale? _____

0   0.5

11 Two angles of a triangle are 54° and 45°.

What size is the third angle? _____°

12 What is the volume of a cuboid measuring 5 cm by 5 cm by 4 cm? _____ cm³

13 Emilia left home at 11.55 a.m. to visit her friend, arriving at her friend's house at 12.28 p.m. How long did the journey take her?

_____ minutes

Questions 14 to 16 refer to this list showing the numbers of peas on 4 plates.

36 34 48 42

14 What is the range? _____

15 What is the median? _____

16 What is the mean? _____

17 The Carroll diagram shows some information about people in a gymnastics club.

What percentage of the members is male? _____%

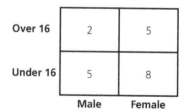

	Male	Female
Over 16	2	5
Under 16	5	8

18 What is the smallest number greater than 60 that is a multiple of both 3 and 7?

_____

19 The product of two integers is 48 and their sum is 19

What is the larger integer? _____

20 A school has 147 pupils. There are 5 more girls than boys.

How many boys are there? _____

Test 33

For all of the questions in this test, do the calculation entirely in your head with no written 'working' and just write down the answer.

In questions 1 to 10 you are reminded of 10 useful strategies which may help you in later questions.

1 21 × 19 _____

  ✱ 20 × 19 or 21 × 20 first

2 55 × 4 _____

  ✱ Double and then double again

3 145 ÷ 5 _____

  ✱ Divide by 10 then multiply by 2

4 128 + 47 _____

  ✱ Add 130 to 47 then subtract 2

5 273 ÷ 21 _____

  ✱ Divide by 3, then by 7

6 12 × 0.3 _____

  ✱ You know that 12 × 3 = 36

7 $\frac{2}{7}$ of 28 _____

  ✱ Find $\frac{1}{7}$ then multiply by 2

8 136 + 58 + 44 _____

  ✱ 136 + 44 first

9 11.1 × 4 _____

  ✱ The result must be about 44

10 13 × 12 _____

  ✱ The units digit of the result is 6

11 How many degrees warmer is 13.5 °C than ⁻3.5 °C?

  _____ degrees

12 Round 16.454 to 3 significant figures.

  _____

13 Write $\frac{6}{5}$ as a mixed number. _____

14 What fraction of the shape is purple?

  _____

15 Given that 8722 ÷ 89 = 98, what is 49 × 89?

  _____

16 What is the remainder when 100 is divided by 8?

  _____

17 By how much is 5 × 12 greater than 2 × 15?

  _____

18 Jakov buys 4 birthday cards priced at £1.95 each and hands over a £10 note. The shopkeeper gives him the change in the smallest number of coins possible.

  How many coins does Jakov receive in change?

  _____

19 Abbie is thinking of a number and gives the following clues:

  "My number is:

  • less than 50
  • a multiple of 3
  • 1 more than a cube number."

  What number is Abbie thinking of?

  _____

20 On this tower of bricks, the number on a brick is the positive **difference** between the numbers on the two supporting bricks.

  What number is on the top brick? _____

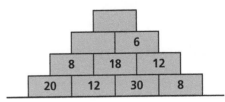

Test 34

For all of the questions in this test, do the calculation entirely in your head with no written 'working' and just write down the answer.

1 Round 30 775 to the nearest 500 _____

2 What is 15% of £500? £ _____

3 Iain and Philippa share a 75 cl bottle of grape juice in the ratio 3:2

   How many centilitres does Philippa drink? _____ cl

4 Jon has the four number cards below.

   What is the largest multiple of 3 he can make by placing cards side by side? _____

5 When Andrea thought of a number, multiplied it by 7 and then subtracted 3, the result was 39

   What number did Andrea think of? _____

6 If $a = 7$ and $b = {}^-4$, what is the value of $4a + 2b$? _____

7 What is the next number in this sequence?

   131, 119, 107, 95, _____

8 Hannah has baked several dozen jam tarts. She sells the tarts at 3 for £1 to raise money for school funds and makes £15 in total. She finds that she has just 3 tarts left and eats them!

   How many dozen tarts did Hannah bake? _____ dozen

9 Calculate the perimeter of the regular hexagon. _____ cm

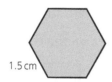

1.5 cm

10 What is the reading on the scale? _____

11 Two angles of a rhombus are each 40°.

   What size is each of the other two angles? _____°

12 A cuboid has volume 60 cm³. It is 5 cm long and 4 cm wide.

   How tall is the cuboid? _____ cm

13 A concert started at 6.25 p.m. and ended $2\frac{1}{2}$ hours later.

   At what time did it end? _____ p.m.

Questions 14 to 16 refer to this list showing the numbers of goals scored in 10 matches.

3 0 5 1 0 4 1 4 0 7

14 What is the mode? _____

15 What is the median? _____

16 What is the mean? _____

17 The Venn diagram shows some information about children cycling to school.

   What percentage of the boys were wearing safety helmets? _____%

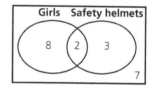

18 Which number is exactly half way between 1000 and 1500? _____

19 The sum of two integers (whole numbers) is 22 and the difference between them is 6

   What is the smaller number? _____

20 In this pictogram, one symbol represents **two** of Ben's model racing cars.

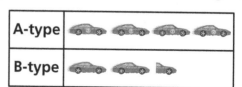

   How many model racing cars does he have?

   _____

Test 35

For all of the questions in this test, do the calculation entirely in your head with no written 'working' and just write down the answer.

In questions 1 to 10 you are reminded of 10 useful strategies which may help you in later questions.

1   17 × 11 _____

   ＊ 17 × 10 first

2   38 × 4 _____

   ＊ Double and then double again

3   85 ÷ 5 _____

   ＊ Divide by 10 then multiply by 2

4   49 + 67 _____

   ＊ Add 50 to 67 then subtract 1

5   575 ÷ 25 _____

   ＊ Divide by 5, then by 5 again

6   12 × 0.9 _____

   ＊ You know that 12 × 9 = 108

7   $\frac{5}{12}$ of 60 _____

   ＊ Find $\frac{1}{12}$ then multiply by 5

8   65 + 39 + 35 + 60 _____

   ＊ 65 + 35 first

9   39 × 5 _____

   ＊ The result must be about 200

10 16 × 7 _____

   ＊ The units digit of the result is 2

11 How many degrees warmer is 5 °C than ⁻6 °C?

   _____ degrees

12 Round 390 645 to 4 significant figures.

   _____

13 Write $\frac{11}{5}$ as a mixed number. _____

14 What proportion of the rectangle is green?

   _____

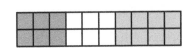

15 Given that 1472 ÷ 46 = 32, what is 32 × 45?

   _____

16 What is the remainder when 72 is divided by 7?

   _____

17 By how much is 4 × 12 greater than 2 × 14?

   _____

18 Marko buys 4 cereal bars priced at 69p each and hands over a £5 note. The shopkeeper gives Marko change using the smallest number of coins possible.

   How many coins does Marko receive?

   _____

19 Leon is thinking of a number and gives the following clues:

   "My number is:

   • between 40 and 70
   • a multiple of 6
   • 1 more than a multiple of 5"

   What number is Leon thinking of?

   _____

20 On this tower of bricks, the number on a brick is the **product** of the numbers on the two bricks supporting it.

   What number is on the brick marked with a star (*)?

   _____

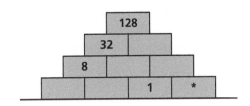

Test 36

For all of the questions in this test, do the calculation entirely in your head with no written 'working' and just write down the answer.

1   Round 2486 to the nearest 100 _____

2   What is 5% of £200?   £ _____

3   John has 6 conkers and Mary has 4 conkers. Write 'the number John has to the number Mary has' as a ratio in its simplest form.

   _____ : _____

4   Omar has the five number cards below.

   What is the largest 2-digit multiple of 7 he can make by placing two of the cards side by side? _____

5   When Lara thought of a number, subtracted 3 and then multiplied by 4, the result was 4

   What number did Lara think of? _____

6   If $a = 1$, $b = {}^{-}1$ and $c = 2$, what is the value of $a - b + c$? _____

7   What is the next number in this sequence?

   1, 3, 11, 43, _____

8   Julia has a tin of buttons. A quarter are black, 16 are white and the other 8 are red.

   How many black buttons are in Julia's tin?

   _____

9   Calculate the area of the triangle.

   _____ cm²

   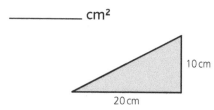

10 What is the reading on the scale? _____

   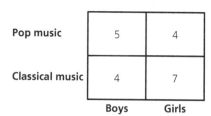

11 Two angles of a parallelogram are each 50°.

   What size is each of the other two angles?

   _____ °

12 What is the volume of a cuboid measuring 2 cm by 2 cm by 14 cm? _____ cm³

13 The walk from school to home takes Robyn 25 minutes. She is expected home from school at 16:15

   At what time does Robyn leave school?

   _____ : _____

Questions 14 to 16 refer to this list showing Barry's 100 m times (in seconds).

15.2   14.8   15.3   14.7   15.0

14 What is the range? _____ seconds

15 What is the median? _____ seconds

16 What is the mean? _____ seconds

17 The Carroll diagram shows the results of a survey into favourite music.

   What percentage of the children prefer classical music? _____ %

	Boys	Girls
Pop music	5	4
Classical music	4	7

18 What is the smallest number greater than 50 that is a multiple of both 3 and 7?

   _____

19 The product of two integers, both less than 10, is 63

   What is the sum of the integers? _____

20 Eric likes dogs. He has twice as many collies as labradors.

   What fraction of his dogs are collies?

   _____

Test 37

For all of the questions in this test, do the calculation entirely in your head with no written 'working' and just write down the answer.

In questions 1 to 10 you are reminded of 10 useful strategies which may help you in later questions.

1   22×19 _____

  * Take easier steps when you multiply

2   22×8 _____

  * Double or halve

3   $230 \div 5$ _____

  * Use your 10s and 2s

4   $249 + 36$ _____

  * Take easier steps when you add

5   $216 \div 12$ _____

  * Use factors to divide

6   8×0.7 _____

  * Use known facts

7   $\frac{2}{5}$ of 65 _____

  * Use a step-by-step approach

8   $117 + 78 + 23$ _____

  * Group when you add

9   40.1×6 _____

  * Approximate the result

10 9×16 _____

  * Check using the units digit of the result

11 How many degrees warmer is 4.9 °C than $^-$1.6 °C?

  _____ degrees

12 Round 16.454 to 3 significant figures.

  _____

13 Write $\frac{9}{2}$ as a mixed number. _____

14 What proportion of the shape is blue?

  _____

15 Given that $3478 \div 47 = 74$, what is 47×37?

  _____

16 What is the remainder when 85 is divided by 7?

  _____

17 By how much is 3^3 greater than 2^3?

  _____

18 Gregory buys 12 birthday candles priced at £1.20 for a pack of 6 and a birthday cake costing £4.25 and hands over a £10 note.

  How much change does he receive? £ _____

19 Andy is thinking of a number and gives the following clues:

  "My number is:
  • between 30 and 70
  • 1 less than a square number
  • 1 more than a prime number."

  What number is Andy thinking of? _____

20 On this tower of bricks, the number on a brick is the positive **difference** between the numbers on the two supporting bricks.

  What number is on the top brick? _____

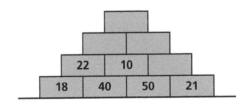

 # Test 38

For all of the questions in this test, do the calculation entirely in your head with no written 'working' and just write down the answer.

1 Round 234 567 to the nearest 1000 _____

2 What is 12% of £50? £ _____

3 Emilia and Florence share a £20 prize. Emilia receives £12

   In what ratio (in its simplest form) do they share the money? _____ : _____

4 Courtney has three number cards.

   She wants to make the largest possible multiple of 3 but 63 is not very large! She can take one extra number card to make the largest possible multiple of 3

   What number card should she choose? _____

5 When Miriam thought of a number, multiplied it by 5 and then added 3, the result was 23

   What number did Miriam think of? _____

6 If $a = 4$ and $b = {}^-1$, what is the value of $3a + 4b$? _____

7 What is the next number in this sequence?

   1, 12, 23, 34, _____

8 Esme bakes 4 apple pies. She cuts each pie into 8 slices and the slices are sold for 80 pence each.

   If she raises £24.00, how many slices remain unsold? _____

9 Calculate the perimeter of the regular star. _____ cm

2 cm

10 What is the reading on the scale? _____

11 Three angles of a kite are 40°, 100° and 100°. What size is the other angle? _____ °

12 A cuboid has volume 54 cm³. It is 3 cm long and 3 cm wide.

   How tall is the cuboid? _____ cm

13 A concert started at 7.15 p.m. and ended 2 hours 50 minutes later.

   At what time did it end? _____ p.m.

Questions 14 to 16 refer to this list showing the numbers of runs scored in 5 innings.

35 0 105 68 12

14 What is the range? _____

15 What is the median? _____

16 What is the mean? _____

17 The Venn diagram shows some information about members in a sports club.

   How many members play just one of tennis and badminton? _____

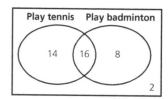

18 Which number is exactly half way between 23 and 75? _____

19 The sum of two integers (whole numbers) is 37 and the difference between them is 11

   What is the smaller number? _____

20 Wilhelm has won stars for good work. In this pictogram, one symbol ☆ represents 1 star. Gold stars earn 5 points each, silver stars 2 points and bronze stars 1 point.

Gold (5)	☆ ☆
Silver (2)	☆ ☆ ☆ ☆
Bronze (1)	☆ ☆ ☆ ☆ ☆ ☆

   How many points has Wilhelm earned?

   _____

Test 39

For all of the questions in this test, do the calculation entirely in your head with no written 'working' and just write down the answer.

In questions 1 to 10 you are reminded of 10 useful strategies which may help you in later questions.

1 45 × 11 _____

   ✱ *Take easier steps when you multiply*

2 396 ÷ 4 _____

   ✱ *Double or halve*

3 34 × 5 _____

   ✱ *Use your 10s and 2s*

4 191 − 56 _____

   ✱ *Take easier steps when you subtract*

5 414 ÷ 18 _____

   ✱ *Use factors to divide*

6 12 × 1.1 _____

   ✱ *Use known facts*

7 $\frac{3}{5}$ of 115 _____

   ✱ *Use a step-by-step approach*

8 18 + 47 + 22 _____

   ✱ *Group when you add*

9 11.1 × 9 _____

   ✱ *Approximate the result*

10 16 × 7 _____

   ✱ *Check using the units digit of the result*

11 How many degrees warmer is 7 °C than ⁻10 °C?

   _____ degrees

12 Round 3599 to 1 significant figure.

   _____

13 Write $\frac{12}{5}$ as a mixed number. _____

14 What fraction of the shape is red?

   _____

15 Given that 1344 ÷ 32 = 42, what is 64 × 42?

   _____

16 What is the remainder when 110 is divided by 9?

   _____

17 By how much is 5 × 14 greater than 4 × 15?

   _____

18 Merida buys 4 chocolate eggs priced at 55p each and 2 bottles of juice priced at 40p each.

   How much change will she receive from a £5 note?

   £ _____

19 Tai is thinking of a number and gives the following clues:

   "My number is:
   • between 40 and 60
   • 1 more than a multiple of 4
   • 2 less than a multiple of 5"

   What number is Tai thinking of? _____

20 On this tower of bricks, the number on a brick is the sum of the numbers on the two bricks supporting it.

   What number is on the brick marked with a star (*)? _____

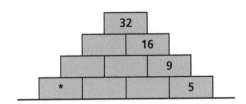

Test 40

For all of the questions in this test, do the calculation entirely in your head with no written 'working' and just write down the answer.

1 Round 44 680 to the nearest 500 _____

2 What is 15% of £1000? £_____

3 The ages of Wendy and Peter are in the ratio 4:7 and Peter is 6 years older than Wendy.

How old is Wendy? _____

4 Johann has the four number cards below.

What is the number closest to 5000 he can make by placing cards side by side?

_____

5 When Nikita thought of a number, subtracted 7 and then multiplied by 2, the result was 18

What number did Nikita think of? _____

6 If $a = 3$ and $b = 2$, what is the value of $2(a + b)$?

_____

7 What is the next number in this sequence?

220, 197, 174, 151, _____

8 Serena has a collection of 25 dolls. 14 are in national costumes.

What percentage of the dolls are not in national costume? _____ %

9 Calculate the area of the triangle.

_____ cm²

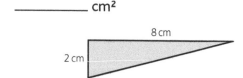

10 What is the reading on the scale? _____

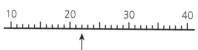

11 Two angles of an isosceles trapezium are each 40°.

What size is each of the other two angles?

_____°

12 What is the volume of a cuboid measuring 5 cm by 10 cm by 10 cm? _____ cm³

13 Agnes left home at 10.55 a.m. and walked to the florists to collect flowers she had ordered. After collecting the flowers, she walked home again at the same speed, arriving at 12.05 p.m.

At what time did Agnes collect the flowers?

_____ a.m/p.m.

Questions 14 to 16 refer to this list showing Mel's five best 200 m times, in seconds.

32.5 32.4 32.0 32.5 32.1

14 What is the range? _____ seconds

15 What is the median? _____ seconds

16 What is the mean? _____ seconds

17 The Carroll diagram shows some information about people attending a dinner.

How many more females were at the dinner than males? _____

	Female	Male
Not vegetarian	6	5
Vegetarian	5	4

18 What is the smallest number greater than 20 that is divisible exactly by 19? _____

19 The product of two integers, both less than 10, is 56

What is the sum of the integers? _____

20 Martin has 12 model aeroplanes. One-sixth of them are jets. How many are not jets?

_____

Test 41

For all of the questions in this test, do the calculation entirely in your head with no written 'working' and just write down the answer.

In questions 1 to 10 you are reminded of 10 useful strategies which may help you in later questions.

1 21×21 _____

   ✱ *Take easier steps when you multiply*

2 123×4 _____

   ✱ *Double or halve*

3 $220 \div 5$ _____

   ✱ *Use your 10s and 2s*

4 $239 + 64$ _____

   ✱ *Take easier steps when you add*

5 $306 \div 18$ _____

   ✱ *Use factors to divide*

6 1.5×0.4 _____

   ✱ *Use known facts*

7 $\frac{2}{9}$ of £6.30 £ _____

   ✱ *Use a step-by-step approach*

8 $49 + 17 + 63 + 21$ _____

   ✱ *Group when you add*

9 £29.90 × 4 £ _____

   ✱ *Approximate the result*

10 14×13 _____

   ✱ *Check using the units digit of the result*

11 How many degrees warmer is 4°C than ⁻1°C?

   _____ degrees

12 Round 304 to 2 significant figures.

   _____

13 Write $\frac{7}{4}$ as a mixed number. _____

14 The diagram shows a piece of card.

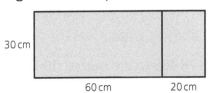

Terry cuts off a piece 20 cm wide as shown.

What proportion of the piece of card does he cut off?

_____

15 Given that $540 \div 36 = 15$, what is 16×36?

_____

16 What is the remainder when 80 is divided by 11? _____

17 By how much is 7×15 greater than 5×17?

_____

18 Vedad buys 2 burgers priced at £3.45 each and 2 cans of cola priced at 75p each.

How much change will he receive from a £10 note? £ _____

19 Kerry is thinking of a number and gives the following clues:

"My number is:
• between 50 and 100
• prime

and the digits add to 10"

What number is Kerry thinking of?

_____

20 On this tower of bricks, the number on a brick is the **product** of the numbers on the two bricks supporting it.

What number is on the brick marked with a star (*)? _____

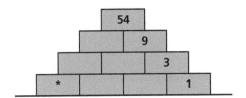

 # Test 42

For all of the questions in this test, do the calculation entirely in your head with no written 'working' and just write down the answer.

1 Round 78 465 to the nearest 1000 _____

2 What is 25% of £18? £ _____

3 There are between 31 and 35 toffees in a bag. When 4 friends share them fairly, there are 2 toffees left over.

How many toffees were in the bag?

_____

4 Rudolf has the three number cards below.

What is the largest common multiple of 3 and 5 he can make by placing cards side by side? _____

5 When Leona thought of a number, added 4 and then multiplied by 3, the result was 45

What number did Leona think of? _____

6 If $a = 4$, $b = ^-2$ and $c = 3$, what is the value of $a + 2b + c$? _____

7 What is the missing number in this sequence?

1, 2, 6, 22, _____, 342

8 Agatha has a collection of detective novels. 3 are hardbacks, 2 are on audio CD and 80% of the collection are paperbacks.

How many detective novels does Agatha have? _____

9 Calculate the area of the triangle.

_____ cm²

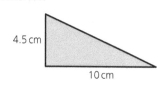

10 What is the reading on the scale? _____

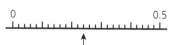

11 Two angles of a triangle are 90° and 55°.

What size is the third angle? _____°

12 A cuboid is 8 cm long and 2.5 cm wide. The volume is 80 cm³.

How tall is the cuboid? _____ cm

13 Wanda left home at 11.50 a.m. to visit her friend, arriving at her friend's house at 12.05 p.m. How long did the journey take her?

_____ minutes

Questions 14 to 16 refer to this list showing the numbers of apples on 4 trees.

10 42 28 30

14 What is the range? _____

15 What is the median? _____

16 What is the mean? _____

17 The Carroll diagram shows some information about people at a football match.

How many people were at the match?

_____

	Male	Female
Over 16	2047	1512
Under 16	1488	953

18 What is the smallest number greater than 60 that is a multiple of both 6 and 9?

_____

19 The product of two integers is 72 and their sum is 17

What is the larger integer? _____

20 A school has 156 pupils. There are 12 more boys than girls.

How many boys are there? _____

Test 43

For all of the questions in this test, do the calculation entirely in your head with no written 'working' and just write down the answer.

In questions 1 to 10 you are reminded of 10 useful strategies which may help you in later questions.

1 21 × 17 _____

   ✳ *Take easier steps when you multiply*

2 48 × 4 _____

   ✳ *Double or halve*

3 295 ÷ 5 _____

   ✳ *Use your 10s and 2s*

4 249 + 153 _____

   ✳ *Take easier steps when you add*

5 228 ÷ 12 _____

   ✳ *Use factors to divide*

6 7 × 0.8 _____

   ✳ *Use known facts*

7 $\frac{2}{5}$ of 405 _____

   ✳ *Use a step-by-step approach*

8 3.45 + 6.93 + 1.55 _____

   ✳ *Group when you add*

9 15.1 × 4 _____

   ✳ *Approximate the result*

10 14 × 13 _____

   ✳ *Check using the units digit of the result*

11 How many degrees warmer is 0.5 °C than ⁻4.5 °C?

   _____ degrees

12 Round 16.75 to 2 significant figures.

   _____

13 Write $\frac{8}{5}$ as a mixed number. _____

14 What proportion of the shape is purple?

   _____

15 Given that 3402 ÷ 63 = 54, what is 108 × 63?

   _____

16 What is the remainder when 110 is divided by 3?

   _____

17 By how much is 6 × 14 greater than 4 × 16?

   _____

18 Ali buys 6 bottles of lemonade priced at £1.45 each and hands over a £10 note. The shopkeeper gives him the change in the smallest number of coins possible.

   How many coins does Ali receive in change?

   _____

19 Ivan is thinking of a number and gives the following clues:

   "My number is:
   • less than 50
   • a multiple of 4
   • 1 more than a cube number."

   What number is Ivan thinking of? _____

20 On this tower of bricks, the number on a brick is the positive **difference** between the numbers on the two supporting bricks.

   What number is on the top brick? _____

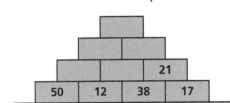

Test 44

For all of the questions in this test, do the calculation entirely in your head with no written 'working' and just write down the answer.

1   Round 409 520 to the nearest 10 000

_____

2   What is 5% of £24?   £ _____

3   Esme and Ben share a box of chocolates in the ratio 4:5

Ben eats 4 more chocolates than Esme. How many chocolates were in the box? _____

4   Hans has the four number cards below.

What is the largest multiple of 6 he can make by placing cards side by side? _____

5   When Gustavus thought of a number, multiplied it by 5 and then subtracted 7, the result was 8

What number did Gustavus think of?

_____

6   If $a = 5$ and $b = {}^-1$, what is the value of $2a + 3b$? _____

7   What is the next number in this sequence?

200, 157, 114, 71, _____

8   Sofie has baked some cupcakes. She sells the cupcakes at 4 for £1.50 to raise money for charity and takes £10.50 in total. She has just 4 cupcakes left.

How many cupcakes did she bake? _____

9   Calculate the perimeter of the regular octagon.   _____ cm

2.5 cm

10 What is the reading on the scale? _____

```
 ⁻2        ⁻1        0         1         2
 |ıııı|ıııı|ıııı|ıııı|ıııı|ıııı|ıııı|ıııı|
      ↑
```

11 Two angles of an isosceles triangle are each 37°.

What size is the other angle? _____°

12 A cuboid has volume 25 cm³. It is 5 cm long and 2 cm wide.

How tall is the cuboid? _____ cm

13 A concert started at 7.25 p.m. and ended $2\frac{3}{4}$ hours later.

At what time did it end? _____ p.m.

Questions 14 to 16 refer to this list showing the numbers of goals scored in 10 matches.

2 1 3 2 4 0 2 3 2 3

14 What is the mode? _____

15 What is the median? _____

16 What is the mean? _____

17 The Venn diagram shows some information about children in a school.

How many of the children do not sing in the choir? _____

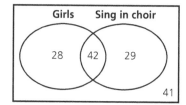

18 Which number is exactly half way between 45 and 83? _____

19 The sum of two integers (whole numbers) is 29 and the difference between them is 7

What is the smaller number? _____

20 In this pictogram, one symbol 🐱 represents **two** cats at a rescue centre?

Female	🐱🐱🐱🐱🐱🐱🐱🐱🐱
Male	🐱🐱🐱🐱🐱

A quarter of the male cats and 9 female cats find new homes.

How many cats remain at the rescue centre?

_____

Test 45

For all of the questions in this test, do the calculation entirely in your head with no written 'working' and just write down the answer.

In questions 1 to 10 you are reminded of 10 useful strategies which may help you in later questions.

1   23 × 11 _____

   ✳ Take easier steps when you multiply

2   22 × 8 _____

   ✳ Double or halve

3   160 × 5 _____

   ✳ Use your 10s and 2s

4   84 + 79 _____

   ✳ Take easier steps when you add

5   256 ÷ 16 _____

   ✳ Use factors to divide

6   11 × 1.2 _____

   ✳ Using known facts

7   $\frac{3}{8}$ of £200   £ _____

   ✳ Use a step-by-step approach

8   73 + 45 + 37 + 5 _____

   ✳ Group when you add

9   21 × 9 _____

   ✳ Approximate the result

10 14 × 6 _____

   ✳ Check using the units digit of the result

11 How many degrees warmer is ⁻6 °C than ⁻11°C?

   _____ degrees

12 Round 38 451 to 3 significant figures.

   _____

13 Write $\frac{17}{5}$ as a mixed number. _____

14 What proportion of the rectangle is purple?

   _____

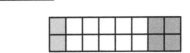

15 Given that 26 × 34 = 884, what is 17 × 13?

   _____

16 What is the remainder when 49 is divided by 6?

   _____

17 By how much is 7 × 13 greater than 3 × 17?

   _____

18 Daiki buys 3 newspapers priced at £1.45 each and hands over a £5 note. The shopkeeper gives Daiki change using the smallest number coins possible.

   How many coins does Daiki receive?

   _____

19 Leonora is thinking of a number and gives the following clues:

   "My number is:

   • between 60 and 100
   • a multiple of 11
   • 1 less than a multiple of 5"

   What number is Leonora thinking of?

   _____

20 On this tower of bricks, the number on a brick is the sum of the numbers on the two bricks supporting it.

   What number is on the brick marked with a star (*)?

   _____

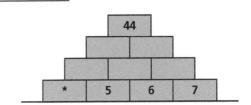

Test 46

For all of the questions in this test, do the calculation entirely in your head with no written 'working' and just write down the answer.

1 Round 75 261 to the nearest 500 _____

2 What is 11% of £120? £ _____

3 Julian has 8 marbles and George has 6 marbles. Write 'the number Julian has to the number George has' as a ratio in its simplest form.

_____ : _____

4 Suzi has the five number cards below.

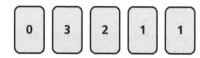

What is the smallest **4-digit** odd number she can make by placing 4 of the cards side by side? _____

5 When Martin thought of a number, subtracted 4 and then multiplied by 5, the result was 30

What number did Martin think of? _____

6 If $a = 2$, $b = {}^-5$ and $c = 1$, what is the value of $3a - b + 2c$? _____

7 What is the next number in this sequence?

1, 4, 10, 22, 46, _____

8 Maya has a bag of sweets. Half are toffees, a third are chocolates and the other 4 are mints.

How many sweets are in the bag? _____

9 Calculate the area of the triangle.
_____ cm²

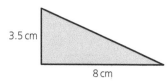

3.5 cm

8 cm

10 What is the reading on the scale? _____

5.0   5.5

11 Two angles of a parallelogram are each 110°.

What size is each of the other two angles?

_____ °

12 What is the volume of a cuboid measuring 3 cm by 4 cm by 5 cm?

_____ cm³

13 Elsa starts a sponsored walk at 09:30 and she finishes $5\frac{3}{4}$ hours later.

At what time does Elsa finish walking?

_____ : _____

Questions 14 to 16 refer to this list showing Barry's 400 m times (in seconds).

71.5 70.5 71.2 71.8 70.0

14 What is the range? _____ seconds

15 What is the median? _____ seconds

16 What is the mean? _____ seconds

17 The Carroll diagram shows the results of a survey about favourite pets.

What percentage of children in the survey was girls? _____ %

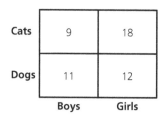

	Boys	Girls
Cats	9	18
Dogs	11	12

18 What is the largest number less than 100 that is a multiple of both 5 and 8?

_____

19 The product of two integers, both less than 10, is 48

What is the sum of the integers? _____

20 Erica is 4 years older than her younger brother Falcon. In 2 years' time, Falcon will be 9

How old is Erica now? _____

Test 47

For all of the questions in this test, do the calculation entirely in your head with no written 'working' and just write down the answer.

In questions 1 to 10 you are reminded of 10 useful strategies which may help you in later questions.

1 21 × 18 _____

   ＊ *Take easier steps when you multiply*

2 73 × 4 _____

   ＊ *Double or halve*

3 470 ÷ 5 _____

   ＊ *Use your 10s and 2s*

4 298 + 57 _____

   ＊ *Take easier steps when you add*

5 364 ÷ 14 _____

   ＊ *Use factors to divide*

6 0.5 × 0.6 _____

   ＊ *Use known facts*

7 $\frac{4}{9}$ of 360 _____

   ＊ *Use a step-by-step approach*

8 £1.95 + £6.05 + £3.55   £ _____

   ＊ *Group when you add*

9 24.9 × 6 _____

   ＊ *Approximate the result*

10 16² _____

   ＊ *Check using the units digit of the result*

11 How many degrees warmer is 0.7 °C than ⁻3.8 °C?

   _____ degrees

12 Round 0.0465 to 2 decimal places.

   _____

13 Write $\frac{13}{4}$ as a mixed number. _____

14 What proportion of the shape is blue?

   _____

15 Given that 3618 ÷ 67 = 54, what is 3618 ÷ 134? _____

16 What is the remainder when 1000 is divided by 3? _____

17 By how much is 8 × 13 greater than 3 × 18?

   _____

18 Lewis buys 4 ice-creams priced at £1.19 each and hands over a £5 note. The shopkeeper gives him the change in the smallest number of coins possible.

   How many coins does Lewis receive in change? _____

19 Dmitri is thinking of a number and gives the following clues:

   "My number is:

   • between 35 and 85
   • a multiple of 12
   • a multiple of 15"

   What number is Dmitri thinking of?

   _____

20 On this tower of bricks, the number on a brick is the **product** of the numbers on the two supporting bricks.

   What number is on the brick marked with a star (*)?

   _____

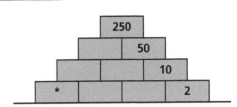

Test 48

For all of the questions in this test, do the calculation entirely in your head with no written 'working' and just write down the answer.

1   Round 461 499 to the nearest 1000 _____

2   What is 2% of £12?   £ _____

3   Lorenzo and Livio share a packet of biscuits in the ratio 4:3

   Lorenzo eats 3 more biscuits than Livio. How many biscuits were in the packet? _____

4   Nils has the four number cards below.

   24 and 123 both have a digit sum of 6

   How many numbers with a digit sum of 6 (including the two examples above) can he make by placing cards side by side? _____

5   When Justine thought of a fraction, multiplied it by 3 and then added 2, the result was 4

   What fraction did Justine think of? _____

6   If $a = 6$ and $b = {}^-3$, what is the value of ab? _____

7   What is the next number in this sequence?

   120, 60, 30, 15, _____

8   Annabel has made 12 pancakes. The ingredients cost £3.00

   She sells the pancakes for 45p each with the profit given to a charity.

   How much money does the charity receive? £ _____

9   Calculate the perimeter of the regular decagon. _____ cm

4.3 cm

10 What is the reading on the scale? _____

$$\begin{array}{ccccc} {}^-4 & {}^-2 & 0 & 2 \end{array}$$
↑

11 Two angles of an isosceles trapezium are each 123°.

   What size is each of the other two angles? _____°

12 A cuboid has volume 240 cm³. It is 8 cm long and 5 cm wide.

   How tall is the cuboid? _____ cm

13 A $2\frac{3}{4}$ hour concert ended at 10.15 p.m.

   At what time did it start? _____ p.m.

Questions 14 to 16 refer to this list showing the numbers of runs scored in 4 innings of cricket.

213 38 27 162

14 What is the range? _____

15 What is the median? _____

16 What is the mean? _____

17 The Venn diagram shows some information about children in a school.

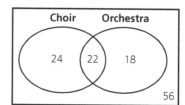

   What fraction of the children is in the orchestra? _____

18 Which number is exactly half way between 37 and 73? _____

19 The sum of two proper fractions is 1 and the difference between them is $\frac{1}{2}$

   What is the smaller fraction? _____

20 In this pictogram, one symbol [🥤] represents **two** cans of drink sold.

| Cola (80p) | 🥤 🥤 🥤 🥤 🥤 |
| Orange (90p) | 🥤 🥤 🥤 🥤 |

   What was the total amount taken? £ _____

Test 49

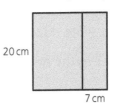
For all of the questions in this test, do the calculation entirely in your head with no written 'working' and just write down the answer.

In questions 1 to 10 you are reminded of 10 useful strategies which may help you in later questions.

1 29 × 22 _____

    ✱ *Take easier steps when you multiply*

2 13.5 × 4 _____

    ✱ *Double or halve*

3 44 ÷ 5 _____

    ✱ *Use your 10s and 2s*

4 4999 + 178 _____

    ✱ *Take easier steps when you add*

5 777 ÷ 21 _____

    ✱ *Use factors to divide*

6 2.5 × 0.2 _____

    ✱ *Use known facts*

7 $\frac{4}{7}$ of £35.70   £ _____

    ✱ *Use a step-by-step approach*

8 176 + 49 + 24 + 131 _____

    ✱ *Group when you add*

9 £59.90 × 4   £ _____

    ✱ *Approximate the result*

10 13 × 23 _____

    ✱ *Check using the units digit of the result*

11 How many degrees cooler is ⁻14 °C than 4 °C?

_____ degrees

12 Round 34 995 to 3 significant figures.

_____

13 Write $4\frac{3}{5}$ as an improper fraction.

_____

14 The diagram shows a square piece of card.

20 cm / 7 cm

Tarik cuts off a piece 7 cm wide as shown.

What proportion of the piece of card does he cut off?

_____

15 Given that 39 × 158 = 6162, what is 158 × 38?

_____

16 What is the remainder when 400 is divided by 6? _____

17 By how much is 9 × 13 greater than 3 × 19?

_____

18 Valerie buys 4 sandwiches priced at £1.35 each and 4 cans of lemonade priced at 85p each.

How much change will she receive from a £20 note?   £ _____

19 Kyle is thinking of a number and gives the following clues:

"My number is:

• between 10 and 60
• prime

and the digits add to 7"

What number is Kyle thinking of? _____

20 On this tower of bricks, the number on a brick is the sum of the numbers on the two bricks supporting it.

What number is on the brick marked with a star (*)? _____

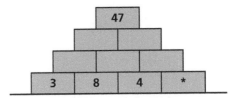

Test 50

For all of the questions in this test, do the calculation entirely in your head with no written 'working' and just write down the answer.

1 Round 145 251 to the nearest 5000 _____

2 What is 5% of £4500? £ _____

3 There are between 40 and 45 sweets in a packet. When 6 friends share them fairly, there are 2 sweets left over.

   How many sweets were in the packet? _____

4 Ronnie has the seven number cards below.

   What is the number nearest to 50 000 he can make by placing cards side by side? _____

5 When Arabella thought of a fraction, added $\frac{1}{2}$ and then multiplied by 4, the result was 5

   What fraction did Arabella think of? _____

6 If $a = \frac{1}{2}$, $b = {}^-2$ and $c = 5$, what is the value of $4a - b + c$? _____

7 What is the missing number in this sequence?

   1, 2, 5, 14, _____, 122

8 Lachlan has a collection of commemorative 50p coins with a face value of £13.50

   Two-thirds of the coins are 2012 Olympic 50p coins.

   How many of his 50p coins commemorate other events? _____

9 Calculate the area of the triangle. _____ cm²

   3.5 cm
   6.4 cm

10 What is the reading on the scale? _____

   45   50
   ↑

11 Three angles of a quadrilateral are 77°, 125° and 111°.

   What size is the fourth angle? _____°

12 A cuboid is 5 cm long and 2 cm wide. The volume is 10 cm³.

   What is the total surface area of the cuboid?

   _____ cm²

13 Mariana left home at 11.40 a.m. to post a letter. After posting the letter, she walked the same route home at the same speed, arriving at 12.16 p.m.

   At what time did she post the letter?

   _____ a.m. /p.m.

Questions 14 to 16 refer to this list showing the numbers of pears on 6 trees.

13 41 29 37 23 37

14 What is the range? _____

15 What is the median? _____

16 What is the mean? _____

17 The Carroll diagram shows some information about people voting in a charity beauty competition.

	Male voters	Female voters
Finalist A	148	72
Finalist B	123	137

   By how many votes did finalist B win?

   _____

18 What is the smallest number greater than 1000 that is a multiple of 6? _____

19 The product of two integers, both less than 20, is 91

   What is their sum? _____

20 A judo club has 35 members. There are 9 more girls than boys.

   How many girls are there? _____